MATERIALS SCIENCE AND TECHNOLOGIES

FUNCTIONS OF TUNGSTEN ALLOYING IN MICROALLOYED STEELS

Materials Science and Technologies

Additional books in this series can be found on Nova's website under the Series tab.

Additional e-books in this series can be found on Nova's website under the e-book tab.

MATERIALS SCIENCE AND TECHNOLOGIES

FUNCTIONS OF TUNGSTEN ALLOYING IN MICROALLOYED STEELS

JINGWEI ZHAO, PH.D.
ZHENGYI JIANG, PH.D.
AND
CHONG SOO LEE, PH.D.

nova publishers
New York

Library of Congress Cataloging-in-Publication Data

ISBN: 978-1-63321-149-0

Published by Nova Science Publishers, Inc. † New York

CONTENTS

PREFACE

Microalloyed steels have extensive applications in metal manufacturing industries and our daily lives. With the progress of science and technology, the demands for microalloyed steels with specified performance being ensured for various service environments have significantly increased. As an alloying element with growing scientific and industrial interests, tungsten has been putting a great attention due to its positive roles in microalloyed steels. Although there have been a lot of literatures on various aspects of tungsten alloying in tool steels, high speed steels, heat resisting steels and stainless steels, only limited published work, however, is available with regard to tungsten alloying in microalloyed steels, let alone a book with respect to this subject. Tungsten alloying in microalloyed steels is still a rear and fruitful research field so far. Such background has prompted and encouraged us to write a book to fulfil such a gap in the area of steel research.

This new book of *Functions of Tungsten Alloying in Microalloyed Steels* is not only a technical document that can be used in the research and development of tungsten alloying in microalloyed steels, but also a scientific and valuable literature for the metallurgists, engineers, materials scientists, academics and higher degree research students who make, use, study and design with steels and tungsten throughout the world. The authors have been intimately and significantly involved in the development of microalloyed steels at the universities in Australia, Republic of Korea and PR China for many years and have established very close relationships with major steel-making companies including BlueScope, POSCO and Baosteel. In this book, the latest research findings and achievements obtained in many aspects of the functions of tungsten alloying in microalloyed steels have been addressed. We have done our best to make this book more accessible to readers, and hope this book will be a turning-point for the development of new microalloyed steels with improved quality and properties to make them applicable to various application environments.

This book considers the metallurgical basis for steel research. Readers are expected to have relevant knowledge of metallurgy, and consequently some introductory explanations are omitted. However, we have included as many references as possible at the end of each chapter to provide readers a way to access relevant knowledge in relating to the contents presented. The text of this book is up-to-date, informative, readable and self-contained. We have attempted to outline the principles of tungsten alloying in microalloyed steels. Many aspects, including the background knowledge of tungsten and microalloyed steels, and the effects of tungsten on the phase transformation, precipitation, microstructural evolution, mechanical properties, optimal heat treatment conditions, corrosion resistance and hydrogen embrittlement, that need to be assessed for microalloyed steels when they are used in practice have been addressed.

In Chapter 1, the development of tungsten and microalloyed steels, and the significance of tungsten alloying in microalloyed steels are described. Chapter 2 states the fundamental issues of phase transformations in steels at first, and then analyses the effect of tungsten on phase transformation. The role of precipitate in microalloyed steels, and the effect of tungsten on the characteristics of complex precipitates are covered in Chapter 3. In Chapter 4, the microstructural evolution, change of pearlite interlamellar spacing and formation of allotriomorphic ferrite of microalloyed steels after tungsten addition is analysed; emphasis is placed on the transition mechanism from acicular ferrite to bainite after tungsten addition. Chapter 5 discusses the variations of room/low temperatures mechanical properties of microalloyed steels with different tungsten additions, and suggests the optimal tungsten content that results in the best combination of excellent room/low temperature tensile strength and ductility. The strategies for heat treatment of microalloyed steels are described in Chapter 6, and the optimal heat treatment conditions for microalloyed steels with different tungsten additions are determined. Chapter 7 considers the corrosion resistance of tungsten-containing microalloyed steels, and the positive effect of tungsten on the corrosion resistance is discussed. Chapter 8 describes the effects of tungsten on the hydrogen embrittlement of microalloyed steels. Conclusion remarks and the problems faced at present and the possibility for future development of tungsten alloying in microalloyed steels are stated in Chapter 9.

Very many people including academic staff, higher degree research students and technicians have helped with the book and it is impossible to acknowledge each by name but their contributions are nevertheless greatly appreciated. One author of Jingwei Zhao is deeply grateful to his wife, Shuang Guo, for her unwavering support and selfless dedication in the taking care of their young daughter, Yuehan Zhao, so that he has sufficient time to consult relevant documents and organise the manuscript of this book.

Jingwei Zhao, Ph.D.
University of Wollongong, Australia

Zhengyi Jiang, Ph.D.
University of Wollongong, Australia

Chong Soo Lee, Ph.D.
Pohang University of Science and Technology, Republic of Korea

May 2014

ABOUT THE AUTHORS

Dr. Jingwei Zhao, Ph.D.
(Corresponding author)
School of Mechanical, Materials and Mechatronic Engineering,
University of Wollongong,
NSW 2522, Australia
E-mail address: jwzhaocn@gmail.com (J. Zhao)

Prof. Zhengyi Jiang, Ph.D.
School of Mechanical, Materials and Mechatronic Engineering,
University of Wollongong, NSW 2522, Australia

Prof. Chong Soo Lee, Ph.D.
Graduate Institute of Ferrous Technology,
Pohang University of Science and Technology,
Pohang 790-784, Republic of Korea

Chapter 1

INTRODUCTION

ABSTRACT

This chapter provides an introductory overview of tungsten and microalloyed steels. The history of tungsten is firstly presented. The properties and uses of tungsten are then discussed. The applications of tungsten alloying in different main steel grades including high speed steels, tool steels, heat resistant steels, corrosion resistant steels and other steels are summarised. This chapter provides basic information of microalloyed steels, and lists the requirements of chemical compositions and mechanical properties under ASTM standard specifications. Alloying elements that are commonly used and their roles in microalloyed steels are discussed. In addition, the applications of microalloyed steels in metal manufacturing industries and our daily lives are described. Summary views are presented on the potential applications and significance of tungsten alloying in microalloyed steels. Finally, the purpose of writing this book is stated, and the organisation of this book is described.

1.1. HISTORICAL SURVEY OF TUNGSTEN

The word tungsten comes from the Swedish words "tung sten" meaning "heavy stone". Its chemical symbol, W, is the abbreviation of "wolfram", named after the tungsten mineral wolframite. In 1781, a Swedish chemist Carl Wilhelm Scheele discovered that an unknown acid, which he called tungstic acid, could be made from scheelite (at the time named tungsten). In the same year, Torbern Bergman, a Swedish chemist and mineralogist, suggested the possibility of obtaining a new metal by charcoal reduction of the obtained acid. In 1783, two Spanish chemists, the brothers Juan José and Fausto Elhuyar were the first to isolate tungsten through the reduction of this acid with charcoal in samples of the mineral wolframite ($(Fe,Mn)WO_4$). For this reason, the brothers are regarded as the discoverers of the element tungsten. Tungsten mainly exists in wolframite, scheelite ($CaWO_4$), huebnerite ($MnWO_4$) and ferberite ($FeWO_4$). Today, tungsten is still primarily obtained from these minerals using the same basic method developed by José and Elhuyar. China is today by far the largest supplier of primary tungsten, which is reported to have about 75% of the world's tungsten resources [1]. The other principal producing countries are Russia, Austria, Bolivia, Canada, Portugal and Thailand.

1.2. PROPERTIES OF TUNGSTEN

Tungsten is a metallic transition element. Its basic chemical information is characterised as [2]:

Chemical Symbol	W
Period in Periodic Table	6
Group in Periodic Table	6
Atomic Number	74
Average Relative Atomic Mass	183.85 ± 0.03

Tungsten neutral atom contains 74 electrons and 110 neutrons. The electron configuration of the unexcited tungsten atom is defined by [Xe] $4f^{14}5d^{4}6s^{2}$, as schematically illustrated in Figure 1.1. The K, L, M and N shells are completed while; the O and P shells are incomplete. The most common formal oxidation state of tungsten is +6, but it exhibits all oxidation states from -2 to +6 [3,4]. Tungsten typically combines with oxygen to form the yellow tungstic oxide, WO_3, which dissolves in aqueous alkaline solutions to form tungstate ions, WO_4^{2-}.

Tungsten and its compounds show generally low toxicity as compared to most of other metals and their compounds. Pure tungsten is a steel-grey to tin-white metal. Ordinary commercial tungsten is brittle and can be worked only with difficulty. Pure tungsten, however, can be cut with a hacksaw, forged, spun, drawn, or extruded. Tungsten objects are also commonly formed by sintering. Tungsten has the highest melting point (3422°C) among all the metals in the Periodic Table, and at temperatures over 1650°C has the highest tensile strength [1]. Tungsten is highly resistant to corrosion and is attacked only by most mineral acids.

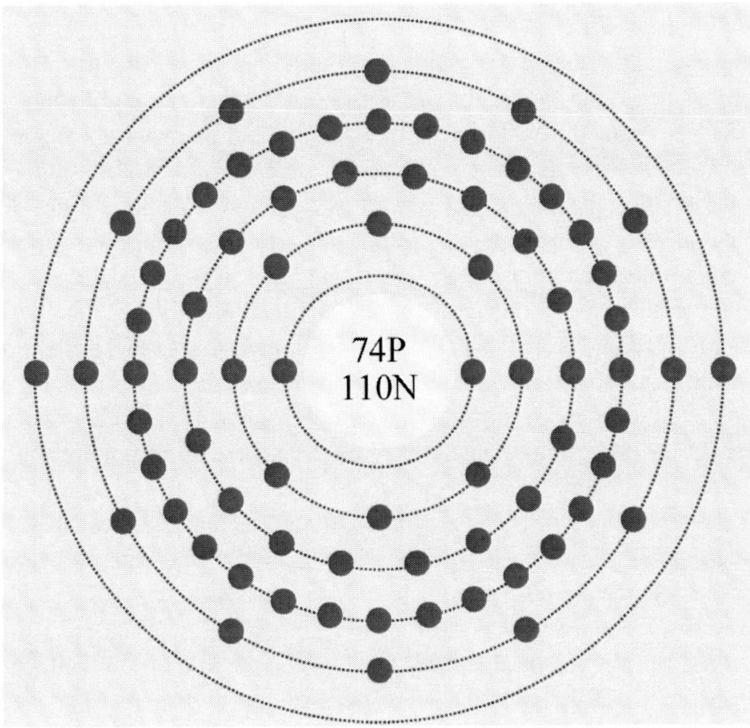

Figure 1.1. Schematic illustration of the atomic structure of tungsten.

It is one of the five major refractory metals (metals with very high resistance to heat and wear) in the body-centred cubic (BCC) family. The other refractory metals are niobium, molybdenum, tantalum and rhenium. Tungsten is also one of the few metals that have a very high mass density (19.25 g/cm^3 at 20°C). Tungsten has the lowest coefficient of thermal expansion of any pure metal. The thermal expansion of tungsten is about the same as borosilicate glass, which makes the metal useful for glass-to-metal seals. The low thermal expansion and high melting point and tensile strength of tungsten originate from strong covalent bonds formed between tungsten atoms by the 5d electrons [2]. Tungsten oxidises in air and must be protected at elevated temperatures.

1.3. Uses of Tungsten

Tungsten features the lowest vapour pressure of all metals, very high moduli of compression and elasticity, very high thermal creep resistance and high thermal and electrical conductivity. Due to these unique attributes, tungsten has few replacements in a majority of its industrial applications. Pure tungsten is used mainly in electrical applications, but its many alloys are used extensively in a wide range of applications such as light bulbs, cathode-ray tube and vacuum tube filaments, as well as heating elements and nozzles on rocket engines. The high melting point also makes tungsten suitable for aerospace and high temperature uses, including heating elements for electrical furnaces, metal evaporation and welding applications, notably in the gas-tungsten arc-welding process. It is also used in electrodes, and in the emitter tips of field emission electron-beam instruments, such as focused ion beam (FIB) and electron microscopes. The metal is also used in X-ray targets. Tungsten is the most important metal for thermo-emission applications, not only because of its high electron emissivity (which is caused by additions of foreign elements) but also because of its high thermal and chemical stability.

Tungsten alloys are used in catalysts, inorganic pigments and tungsten disulphide high temperature lubricants which are stable to 500°C. Tungsten carbide possesses the largest portion of use among tungsten alloys. Tungsten carbide (WC or W_2C) is produced by heating powdered tungsten with carbon, and is one of the hardest substances in existence with a melting point of 2770°C for WC, and 2780°C for W_2C. WC is an efficient electrical conductor, but W_2C is less so. Tungsten carbide behaves in a manner very similar to that of unalloyed tungsten and is resistant to chemical attack, although it reacts strongly with chlorine to form tungsten hexachloride (WCl_6). Tungsten carbide is widely used to make wear-resistant abrasives and cutters and knives for drills, circular saws, milling and turning tools used by the metalworking, woodworking, mining, petroleum and construction industries [5]. In these applications, tungsten carbide may be combined with cobalt, or coated with titanium nitride or titanium carbide. The jewellery industry makes rings of sintered tungsten carbide, tungsten carbide/metal composites, and also metallic tungsten using nickel as a binding element in the place of cobalt, which is used as a binder for industrial purposes. Sometimes manufacturers or retailers refer to tungsten carbide as a metal, but it is actually a ceramic. Because of the super high hardness of tungsten carbide, rings made of this material are extremely abrasion resistant, and will hold a burnished finish longer than rings made of metallic tungsten. Tungsten carbide rings are brittle, however, and may crack under a sharp blow [6]. Figure 1.2 shows an image of a real ring that is made from tungsten carbide. Besides tungsten carbide, tungsten also finds use in heavy metal alloys that are used in armament, heat sinks and high-density applications, such as weights, counterweights, ballast keels for yachts, tail ballast for commercial aircraft and ballast in racing cars.

Figure 1.2. A ring made from tungsten carbide.

Calcium and magnesium tungstates are widely used in fluorescent lighting; other salts of tungsten are used in the chemical and tanning industries. Tungsten disulphide is a dry, high-temperature lubricant, stable to 500°C. Tungsten bronzes and other tungsten compounds are used in paints. Superalloys that contain tungsten are used in turbine blades and wear-resistant parts and coatings. Tungsten powder is used as a filler material in plastic composites, which are used as a non-toxic substitute for lead in bullets, shot and radiation shields.

1.4. Applications of Tungsten in Steels

In the middle of the 19th century, tungsten was first introduced to technical applications, mainly in steel production. From then until the first quarter of the 20th, its importance in this field steadily increased with the steel industry, and steel became the biggest tungsten consumer. Tungsten is usually used as an alloying element in steels to modify the microstructure and improve the properties including corrosion resistance, workability and mechanical properties etc. Different tungsten-containing steels were developed in Austria, Germany, France and England followed by high speed steels in USA [7].

The applications of tungsten alloying in different main steel grades are summarised as follows:

- **High speed steels.** High speed steels are often simply called tungsten steels which consume most of the tungsten used in steelmaking. The term is descriptive of their ability when cutting metals at the "high speed" in use. High speed steels are high-performance special steels offering high working hardness at temperatures up to 500°C, high wear resistance, high resistance to the softening effect of heat and excellent toughness for effective use in industrial cutting operations due to existence of alloying elements like tungsten, molybdenum, vanadium and chromium which are able to form stable carbides. Table 1.1 presents the chemical compositions of high speed steels [8], indicating that tungsten alloying plays an important role in almost all the high speed steels. High speed steels are mainly used in the manufacture of various cutting tools including drills, milling cutters, tool bites, gear cutters, saw blades and so on.
- **Tool steels.** Tool steels refer to a variety of alloy steels that are used to manufacture tools, dies, and moulds that shape, form and cut other materials including steels, nonferrous metals and plastics. Tool steels derive their excellent properties from their distinctive hardness, resistance to abrasion, ability to hold a cutting edge, and/or resistance to deformation at elevated temperatures. For the most widely used tool steels, the following groups are identified by designations in the American Iron and Steel Institute (AISI) classification system [9]:

  Water-hardening tool steels
  Shock-resisting tool steels
  Oil-hardening cold-work tool steels
  Air-hardening, medium-alloy cold-work tool steels
  High-carbon, high-chromium cold-work tool steels
  Low-alloy special-purpose tool steels
  Mould tool steels
  Chromium hot-work tool steels
  Tungsten hot-work tool steels
  Molybdenum hot-work tool steels
  Tungsten high-speed tool steels
  Molybdenum high-speed tool steels
  Ultrahard high-speed tool steels

Table 1.1. Chemical compositions of high speed steels*

Grade	C	Mn	Si	Cr	V	W	Mo	Co
T1	0.65~0.80	0.10~0.40	0.20~0.40	3.75~4.50	0.90~1.30	17.25~18.75	-	-
T2	0.80~0.90	0.20~0.40	0.20~0.40	3.75~4.50	1.80~2.40	17.50~19.00	1.00 max	-
T4	0.70~0.80	0.10~0.40	0.20~0.40	3.75~4.50	0.80~1.20	17.50~19.00	0.40~1.00	4.25~5.75
T5	0.75~0.85	0.20~0.40	0.20~0.40	3.75~5.00	1.80~2.40	17.50~19.00	0.50~1.25	7.00~9.50
T6	0.75~0.85	0.20~0.40	0.20~0.40	4.00~4.75	1.50~2.10	18.50~21.00	0.40~1.00	11.00~13.00
T8	0.75~0.85	0.20~0.40	0.20~0.40	3.75~4.50	1.80~2.40	13.25~14.75	0.40~1.00	4.25~5.75
T15	1.50~1.60	0.15~0.40	0.15~0.40	3.75~5.00	4.50~5.25	11.75~13.00	1.00 max	4.75~5.25
M1	0.78~0.88	0.15~0.40	0.20~0.50	3.50~4.50	1.00~1.35	1.40~2.10	8.20~9.20	-
M2	0.78~0.88 / 0.95~1.05	0.15~0.40	0.20~0.45	3.75~4.50	1.75~2.20	5.50~6.75	4.50~5.50	-
M3:1	1.00~1.10	0.15~0.40	0.20~0.45	3.75~4.50	2.25~2.75	5.00~6.75	4.75~6.50	-
M3:2	1.15~1.25	0.15~0.40	0.20~0.45	3.75~4.50	2.75~3.25	5.00~6.75	4.75~6.50	-
M4	1.25~1.40	0.15~0.40	0.20~0.45	3.75~4.75	3.75~4.50	5.25~6.50	4.25~5.50	-
M6	0.75~0.85	0.15~0.40	0.20~0.45	3.75~4.50	1.30~1.70	3.75~4.75	4.50~5.50	11.00~13.00
M7	0.97~1.05	0.15~0.40	0.20~0.55	3.50~4.00	1.75~2.25	1.40~2.10	8.20~9.20	-
M10	0.84~0.94 / 0.95~1.05	0.10~0.40	0.20~0.45	3.75~4.50	1.80~2.20	-	7.75~8.50	-
M30	0.75~0.85	0.15~0.40	0.20~0.45	3.50~4.25	1.00~1.40	1.30~2.30	7.75~9.00	4.50~5.50
M33	0.85~0.92	0.15~0.40	0.15~0.50	3.50~4.00	1.00~1.35	1.30~2.10	9.00~10.00	7.75~8.75
M34	0.85~0.92	0.15~0.40	0.20~0.45	3.50~4.00	1.90~2.30	1.40~2.10	7.75~9.20	7.75~8.75
M36	0.80~0.90	0.15~0.40	0.20~0.45	3.75~4.50	1.75~2.25	5.50~6.50	4.50~5.50	7.75~8.75
M41	1.05~1.15	0.20~0.60	0.15~0.50	3.75~4.50	1.75~2.25	6.25~7.00	3.25~4.25	4.75~5.75
M42	1.05~1.15	0.15~0.40	0.15~0.65	3.50~4.25	0.95~1.35	1.15~1.85	9.00~10.00	7.75~8.75
M43	1.15~1.25	0.20~0.40	0.15~0.65	3.50~4.25	1.50~1.75	2.25~3.00	7.50~8.50	7.75~8.75
M44	1.10~1.20	0.20~0.40	0.30~0.55	4.00~4.75	1.85~2.20	5.00~5.75	6.00~7.00	11.00~12.25
M46	1.22~1.30	0.20~0.40	0.40~0.65	3.70~4.20	3.00~3.30	1.90~2.20	8.00~8.50	7.80~8.80
M47	1.05~1.15	0.15~0.40	0.20~0.45	3.50~4.00	1.15~1.35	1.30~1.80	9.25~10.00	4.75~5.25
M48**	1.42~1.52	0.15~0.40	0.15~0.40	3.50~4.00	2.75~3.25	9.50~10.50	4.75~5.50	8.00~10.00
M50	0.78~0.88	0.15~0.45	0.20~0.60	3.75~4.50	0.80~1.25	-	3.90~4.75	-
M52	0.85~0.95	0.15~0.45	0.20~0.60	3.50~4.30	1.65~2.25	0.75~1.50	4.00~4.90	-
M62**	1.25~1.35	0.15~0.40	0.15~0.40	3.50~4.00	1.80~2.10	5.75~6.50	10.00~11.00	-

* Contents of sulphur and phosphorus: $S \leq 0.03$, and $P \leq 0.03$. ** Content of sulphur: $S \leq 0.07$.

Tools steels are complex alloy steels that contain large quantities of alloying elements including tungsten, molybdenum, vanadium, chromium, manganese and cobalt. Tungsten and other alloying elements like molybdenum, vanadium and chromium are strong carbide forming elements, and are of significance for the improvement of mechanical properties in order to meet the increasing service demands of tool steels, since the formation of large volume fractions of carbides formed from these elements provides high hardness, excellent wear resistance and good dimensional control during the applied heat treatment of tool steels [10].

- **Heat resistant steels.** Heat resistant steels are a group of solid solution strengthened alloy steels that used in the production of machinery and equipment working at temperatures over 500°C and limited in the extreme to 750°C [11]. The most important alloying elements in heat resistant steels are chromium for oxidation resistance and nickel for strength and ductility. Other alloying elements including manganese, molybdenum, tungsten, titanium, vanadium, niobium, aluminium, cobalt, zirconium, copper, boron, and the rare earth elements such as cerium, lanthanum and yttrium are usually added in different combinations to improve the steels integrative properties at elevated temperatures. The addition of proper amount of tungsten in heat resistant steels benefits the enhancement of high temperature tensile strength, fracture toughness, creep and fatigue resistance, and long-term serving stability [12-14].
- **Stainless steels.** Stainless steels are a large group of special alloy steels developed primarily to withstand corrosion. Other desirable features may include excellent formability, high room-temperature and cryogenic toughness, and good resistance to scaling, oxidation and creep at elevated temperatures [15]. Chromium alloying with a minimum content of 12% by weight is required to impart corrosion resistance to stainless steels, and many other alloying elements including nickel, molybdenum, tungsten, manganese, silicon and copper etc. may be added to stabilise other phases, provide additional corrosion resistance, and/or produce enhanced mechanical properties. Tungsten has a beneficial effect in improving the passivity of stainless steels and thus increasing the corrosion resistance due to the interaction with the positive ions and of the tungstate inhibitive effect [16]. Tungsten alloying in stainless steels also helps to improve the

creep resistance, strength and stability when serving at elevated temperatures.

- **Other steels.** Besides the aforementioned steels, tungsten has also found applications in structural steels for fusion reactors [17,18], valve steels for automotive engines [19], heat-abrasion resistant steels for burner nozzles [20], sag resistant steels for springs [21], maraging steels for engine components [22,23] and high Co-Ni steels for aircrafts [24] etc. Tungsten alloying in these steels plays a role in improving creep strength, high temperature corrosion resistance, hardness, wear resistance, heat-abrasion resistance, sag resistance and/or tensile strength.

1.5. MICROALLOYED STEELS

1.5.1. What are Microalloyed Steels?

Microalloyed steels were originally called high strength low alloy (HSLA) steels. The term "microalloyed steel" refers to a group of alloy steel that contains small amounts of various alloying elements. The chemical compositions of a specific microalloyed steel may vary for different product thicknesses to meet different property requirements. Microalloyed steels in sheet or plate form typically contains 0.05 to 0.25 wt.% C, and up to 2 wt.% manganese [25]. Individual element is usually bellow 0.1 wt.%, and total alloying elements are generally less than 0.15 wt.%. The common used alloying elements, which serve to improve the physical and working characteristics of microalloyed steel, include vanadium, niobium, titanium, molybdenum, zirconium, copper, boron, and rare-earth metals [26-35]. These alloying elements are usually used in various combinations to benefit steel's properties such as weldability, hardness, strength, toughness, formability, atmospheric corrosion resistance and resistance to fatigue and wear.

1.5.2. Classifications of Microalloyed Steels

Microalloyed steels can be simply classified into six groups based on the alloying elements additions and special processing technologies:

- **Weathering steels.** Small amounts of alloying elements such as copper and phosphorus are added in the steels to improve atmospheric corrosion resistance and solid solution strengthening. The atmospheric corrosion resistance of these steels in most environments is substantially better than that of carbon structural steels. When properly exposed to the atmosphere, these steels can be used bare (unpainted) for many applications.
- **Ferrite-pearlite steels.** These steels contain small amount of strong carbide/carbonitride forming elements for precipitation strengthening, grain refinement and in some instances phase transformation temperature control. The microstructure of these steels comprises of ferrite crystal grains and pearlite blocks. A pearlite block contains some pearlite colonies characterised by the lamellar structures of ferrite and cementite. The hierarchical heterogeneity of pearlite plays a key role in the strength and deformation characteristics of ferrite-pearlite steels. The additions of strong carbide/conbonitride forming elements benefit precipitation strengthening and grain refining.
- **Control-rolled steels.** This group of steels may include hot- and cold-rolled steels with improved strength, toughness, formability, weldability, solution hardening and bake hardenability.
- **Acicular ferrite steels.** These steels are generally produced by quenching followed by tempering. The steels have fine lath-like ferrite grain size with a high dislocation density. The acicular ferrite forms at higher temperatures than upper bainite. They contain very low carbon (less than 0.05% C), and possess excellent combination of high yield strength, weldability, formability and good toughness.
- **Dual-phase steels.** The microstructure of these steels consists of ferrite matrix and dispersed martensite islands. Dual-phase steels have an excellent combination of good ductility and high strength. Other characteristics of dual-phase steels include the absence of a discontinuous yield point and a high strain hardening rate.
- **Inclusion-shape-controlled steels.** These steels provide improved ductility and through-thickness toughness by the small additions of calcium, zirconium, or titanium, or perhaps rare earth elements so that the shape of the sulphide inclusions is changed from elongated stringers to small, dispersed, and almost spherical globules.

It should be noted that these groups are not necessarily distinct classifications since a microalloyed steel may have different characteristics from more than one group.

1.5.3. Chemical Compositions of Microalloyed Steels

Table 1.2 presents the chemical compositions of some microalloyed steels under ASTM standard specifications [36-43]. For a designated microalloyed steel, the chemical compositions shall conform to the corresponding chemical requirements.

1.5.4. Mechanical Properties of Microalloyed Steels

The designed microalloyed steel as represented by the tested specimens shall conform to the requirements for mechanical properties. Table 1.3 lists the tensile requirements of some microalloyed steels under ASTM standard specifications [36-43].

1.5.5. Roles of Alloying Elements in Microalloyed Steels

Microalloyed steels are among one of the most commonly used alloyed steels. An important feature of microalloyed steels is the flexible use of alloying elements in specific combinations to improve the properties of steel significantly. The main alloying elements and their roles in microalloyed steels are described below:

- **Carbon.** The presence of carbon in iron is necessary to make steel. It is the primary hardening element in all steels, and is essential to the formation of cementite and other carbides, pearlite, bainite and iron-carbon martensite. Carbon has a major effect on the properties of microalloyed steels. Carbon is a strong austenite former and strongly promotes an austenitic structure. It benefits the increase of strength, hardness and hardenability. Carbon is the single most important alloying element in all steels.
- **Manganese.** Manganese is normally present in all steels and functions as a deoxidiser. It is added in microalloyed steels to improve the

working performance, and increase the strength, hardenability and wear resistance. Increasing the manganese content however decreases ductility and weldability. Manganese slightly increases the strength of ferrite. It increases the hardness penetration of microalloyed steels in quenching by decreasing the critical queening speed, and thus makes the steels more stable in quenching. Manganese benefits the formation of stable carbides, and the stabilisation of austenite.

- **Silicon.** Silicon is one of the principal deoxidisers used in steelmaking. It has a beneficial effect upon the strength and improves the hardenability of microalloyed steels, and helps to increase the toughness when used in combination with certain other alloying elements. Silicon stabilises ferrite and lowers carbon's solubility in austenite. It narrows the temperature range in which austenite exists. Silicon increases the corrosion and oxidation resistance. It however encourages brittleness, and could resent cracking problems due to the resulting decrease in ductility with the increase of silicon content in microalloyed steels.
- **Chromium.** When added to microalloyed steels, chromium can increase the response to heat treatment, hardenability, strength, toughness, and wear and corrosion resistance. Its presence in microalloyed steels, however, could cause excessive hardness and cracking forging and welds. Chromium is a strong carbide former. Complex chromium-iron carbides dissolve in austenite slowly, and a sufficient heating time must be allowed for prior to quenching. As with manganese in microalloyed steels, chromium has a tendency to increase the hardness penetration.
- **Nickel.** Nickel increases the strength, toughness and hardenability but is ineffective in increasing hardness of microalloyed steels. It also tends to help reduce distortion and cracking during quenching. Nickel does not form carbides in steel. It remains in ferrite, strengthening and toughening the ferrite phase. Nickel encourages grain refinement, and is usually used with chromium to advantage in microalloyed steels. Nickel also improves resistance to oxidation and corrosion. It increases toughness at low temperatures when added in small amounts to microalloyed steels.
- **Copper.** Copper is an austenite former. It lowers the austenite/ferrite transformation temperatures, but insufficiently to produce martensite by air cooling. Copper enhances the corrosion resistance in certain acids, and is beneficial to atmospheric corrosion resistance when

presents in amount exceeding 0.2%. Weathering microalloyed steels usually contain copper in amounts of greater than 0.2%. Copper is a solid solution strengthener, and offers precipitation strengthening in microalloyed steels. Copper additions to microalloyed steels have a strong effect on strength and microstructural evolution.

- **Aluminium.** Aluminium is a deoxidiser, and has been traditionally used for more than 100 years as a deoxidising element in steels. It forms aluminium oxide and decreases the amount of oxygen in the steel and facilitates the production of killed steels. Aluminium in microalloyed steels can combine with nitrogen to form aluminium nitride. Aluminium nitride does not contribute greatly to precipitation hardening, it is however powerful in the grain refinement by inhibiting grain growth, leading to improved mechanical properties.
- **Phosphorus.** Phosphorus in microalloyed steels can have beneficial as well as harmful effects. It is one of the most potent solid-solution strengtheners of ferrite. Phosphorus increases strength, hardness and resistance to atmospheric corrosion, but it decreases ductility and toughness when added to microalloyed steels. Phosphorus additions are also known to increase the tendency to cracking during welding. Phosphorus is normally controlled to low levels, and almost all the microalloyed steels are required to have a phosphorus amount of no greater than 0.04%.
- **Sulphur.** Sulphur is usually regarded as an impurity in microalloyed steels. It decreases ductility, toughness and weldability. The content of sulphur in microalloyed steels is usually limited to 0.05% or less.
- **Vanadium.** Vanadium is a strong carbide/carbonitride forming element. It is easily dissolved in austenite, and vanadium carbide/carbonitride precipitates readily and predictably in ferrite during cooling resulting in grain refinement and precipitation strengthening to microalloyed steels. The presence of vanadium in steel raises the temperature at which grain coarsening sets in and under certain conditions increases the hardenability. The addition of small amounts of vanadium can significantly increase the strength of microalloyed steels. Vanadium has relatively little effect on steelmaking and hot working operations such as casting and rolling due to the high solubility of vanadium carbide. In conjunction with molybdenum, vanadium has an effect in enhancing the creep resistance.

- **Niobium.** Niobium is both a strong ferrite and carbide/carbonitrede former. Niobium is less soluble than vanadium in austenite, and its main contributions in microalloyed steels are to form precipitates at a higher temperature enabling the retardation of austenite recrystallisation during hot working such as rolling and forging, thus promoting a fine-grain structure during cooling. The precipitation strengthening potential of niobium in ferrite is reduced by early precipitation in austenite. The addition of small amounts of niobium can significantly increase the yield strength of microalloyed steels. Niobium microalloyed steels often feature excellent toughness due to the fact that grain refinement is the dominant contribution to the increased yield stress. Due to the low solubility of niobium carbide/carbonitride in austenite, niobium is very effective in microalloyed steels containing low levels of carbon and nitrogen. Niobium retards tempering and can decrease the hardenability of microalloyed steels because it forms very stable precipitates.
- **Titanium.** Titanium is a strong ferrite and a strong carbide/carbonitride former. The principal use of titanium in microalloyed steels is to stabilise precipitates by forming titanium carbide/carbonitride. Titanium nitride (TiN) has the greatest stability and will form initially above the iron liquidus temperature and then if the titanium additions are sufficient, $Ti_4C_2S_2$ will be formed, followed by TiC and TiS. The presence of stable and hard titanium precipitates in microalloyed steels can retard austenite grain coarsening during reheating. Titanium addition in microalloyed steels is very effective in grain refinement and precipitation strengthening. It is also used in microalloyed steels to inhibit the heat-affected zone (HAZ) grain coarsening, and hence improves impact toughness of the HAZ formed in welding.
- **Molybdenum.** Molybdenum is a carbide former. It forms stable carbides, and raises the temperature at which softening takes place on tempering and increases resistance to creep of microalloyed steels at elevated temperatures. When added in microalloyed steels, molybdenum increases the hardenability, slows the critical quenching speed, and improves the high temperature strength and hardness. Molybdenum may produce secondary hardening during the tempering of quenched steels. It enhances the creep strength of microalloyed steels at elevated temperatures. The presence of molybdenum in

microalloyed steels also benefits the improvement of corrosion resistance and temper brittleness.

- **Boron.** A small amount of boron improves efficiently the hardenability of microalloyed steels because it delays the nucleation of ferrite on the transformation from austenite to ferrite due to its segregation at austenite grain boundary. Boron can effectively suppress pearlite and quasi-polygonal ferrite, and broaden the cooling rate region for bainite, ferrite and martensite in deformed microalloyed steels. Its effectiveness is most noticeable at lower carbon levels. The addition of boron is usually in very small amounts ranging from 0.0005 to 0.003%.
- **Zirconium.** Zirconium, similarly to titanium, is a strong nitride and a strong oxide forming element. Zirconium nitride is formed above the iron liquidus temperature when zirconium is added in microalloyed steels since zirconium nitride is a high melting compound and exhibits a very strong affinity for nitrogen. Zirconium nitride can retard austenite grain growth during the reheating process, and contributes to the enhanced impact toughness of the HAZ formed in welding.

1.5.6. Applications of Microalloyed Steels

Microalloyed steels have extensive applications in metal manufacturing industries and our daily lives. They are used in every major steel market sector in many parts of the world and their development has played an important role in the expansion of certain key industries such as oil and gas extraction, construction, and transportation.

The wide applications of microalloyed steels include oil and gas pipelines, heavy-duty highway and off-road vehicles, construction and farm machinery, industrial equipment, storage tanks, mine and railroad cars, barges and dredges, snowmobiles, lawn mowers and passenger car components. Microalloyed steels can also be found applications in bridges, suspension components, heavy equipment, offshore structures, power transmission towers, lighting poles, and building beams and panels etc. Figure 1.3 illustrates the main applications of microalloyed steels.

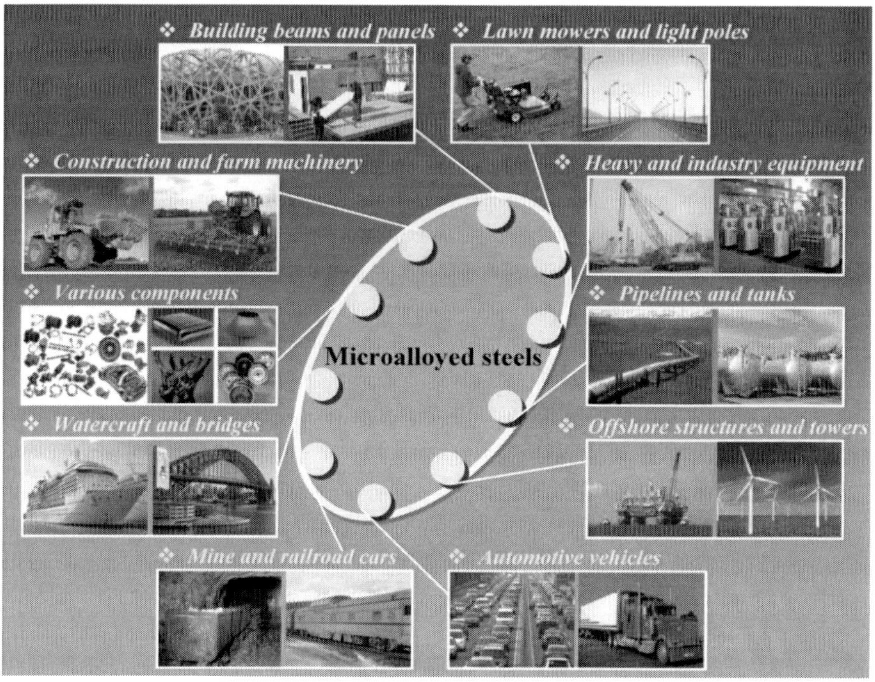

Figure 1.3. Applications of microalloyed steels.

1.6. SIGNIFICANCE OF TUNGSTEN ALLOYING IN MICROALLOYED STEELS

Tungsten is a strong carbide former. When added in microalloyed steels, tungsten may form hard and stable carbides or complex carbides with other carbide forming elements such as vanadium, niobium, titanium, chromium and molybdenum. The addition of tungsten in microalloyed steels will increase the amount of undissolved and excess carbides, and the eutectoid point will be shifted towards lower carbon concentrations. As a result, fine carbides will precipitate and distribute uniformly in the steel matrix. The precipitation of fine and stable carbides in microalloyed steels is very effective for grain size refining and precipitation hardening.

Tungsten decreases the temperature range in which austenite exists. The presence of tungsten in microalloyed steels will raise the austenite/ferrite phase transformation temperature, and the precipitated carbides dissolve slowly over a range of temperature. When completely dissolved in microalloyed steels, tungsten renders transformations sluggish, especially to tempering. It hence increases tempering temperatures.

The addition of a certain amount of tungsten in microalloyed steels will produce a large carbide volume than other alloying elements at the same carbon level.

The wear resistance of microalloyed steels will be enhanced with little or no loss of toughness due to the presence of increased amount of fine and stable carbides. In addition, tungsten addition in microalloyed steels may increase the strength and stability at normal, and low and elevated temperatures. Also when added in microalloyed steels, tungsten may produce fewer tendencies to decarburisation during working, and increase corrosion- and heat-resistance.

As an alloying element with growing scientific and industrial interests, tungsten alloying has been putting great attention due to its positive roles in microalloyed steels. Although there have been a lot of publications on various aspects of tungsten alloying in high speed steels, tool steels, heat resistant steels, stainless steels and other steels, only limited published work is available with regard to tungsten alloying in microalloyed steels, let alone a book with a systematic investigation on the functions of tungsten alloying in these steels. It is hence a rear and fruitful research field, and will be of great significance to write a book in this field by addressing the latest research findings and achievements.

The publication of this book will stimulate the application of tungsten as an alloying element in microalloyed steels, and then promote the rapid development of steel and tungsten companies.

1.7. Purpose and Organisation of This Book

The purpose of writing this book is to provide essential and valuable information to the metallurgists, engineers, materials scientists, academics and graduate students who make, use, study and design with steels and tungsten throughout the world by introducing the latest research findings and achievements of tungsten alloying in microalloyed steels, and considering the metallurgical basis for steel research.

Microalloyed steels are primarily produced by forging and rolling. In this book, several microalloyed steels with a varying tungsten contents that are produced by controlled hot forging process are used as the studying materials.

The chemical compositions of the microalloyed steels are listed in Table 1.4. Ingots were homogenised at 1230°C for 1 h and then hot forged into cuboids with section size of 140 mm × 140 mm and height of 400 mm, followed by air cooling to room temperature. The final forging temperature was controlled to be higher than 950°C. All the specimens for tests were prepared along the height direction of the forged pieces. In this book, the terms "0W", "0.1W", "0.5W", "1W" and "2W" presented in Table 1.4 represent respectively the microalloyed forging steels with 0, 0.1%, 0.5%, 1% and 2% tungsten additions, unless otherwise specified. For example, "0W cast steel" means the microalloyed cast steel without tungsten addition, and its chemical compositions are the same as the 0W forging steel.

Steelmaking and rolling of microalloyed steels are outside the scope of this book even though they are important processing technologies in the production of large-scale high quality microalloyed steels. This book focuses on the functions of tungsten alloying in microalloyed forging steels, especially the effects of tungsten addition on the phase transformation, precipitation characteristics, microstructural evolution, mechanical properties, corrosion resistance and hydrogen embrittlement. To help optimise the microstructure and improve the mechanical properties, heat treatment technologies of tungsten-containing microalloyed steels are also introduced.

Phase transformations will be involved during hot forging and/or subsequent cooling processes.

This book first describes the effect of tungsten on the phase transformation behaviour of microalloyed steels by establishing the continuous cooling transformation (CCT) diagrams.

CCT data are essential for application to most practical hot forging and rolling processes in order to control the expected microstructure and the required mechanical properties. In microalloyed steels, the austenite grain growth behaviour is closely related to the characteristics of precipitates. Fine precipitate dispersions retard austenite growth, and the more stable the precipitates, the more effectively grain growth is retarded to higher temperatures.

Austenite is the parent phase of all the microstructures including pearlite, proeutectoid phases, bainite, matensite and various ferritic microstructures, and plays a key role in the final microstructure.

Therefore, emphasis will then be placed on the effect of tungsten on the characteristics of precipitated particles in microalloyed steels.

Table 1.2. Chemical compositions of microalloyed steels

ASTM specification	Designation	Section thickness (mm)	Chemical compositions (wt.%, max.; unless otherwise specified)								
			C	Mn	Si	P	S	Cr	Ni	Cu	Others
A 242	Type 1	$t \leq 100$	0.15	1.00	---	0.15	0.05	---	---	0.20 min.	---
A 572	Grade 42	$t \leq 150$	0.21	1.35^{A}	0.40^{B}	0.04	0.05	---	---	0.20 min.^{C}	*D*
	Grade 50	$t \leq 100$	0.23	1.35^{A}	0.40^{B}	0.04	0.05	---	---	0.20 min.^{C}	*D*
	Grade 55	$t \leq 50$	0.25	1.35^{A}	0.40^{B}	0.04	0.05	---	---	0.20 min.^{C}	*D*
	Grade 60	$t \leq 32$	0.26	1.35^{A}	0.40	0.04	0.05	---	---	0.20 min.^{C}	*D*
	Grade 65	$13< t \leq 32$	0.23	1.65^{A}	0.40	0.04	0.05	---	---	0.20 min.^{C}	*D*
		$t \leq 13$	0.26	1.35^{A}	0.40	0.04	0.05	---	---	0.20 min.^{C}	*D*
A 588	Grade A	$t \leq 200$	0.19	0.80-1.25	0.30-0.65	0.04	0.05	0.40-0.65	0.40	0.25-0.40	V: 0.02-0.10
	Grade B	$t \leq 200$	0.20	0.75-1.35	0.15-0.50	0.04	0.05	0.40-0.70	0.50	0.20-0.40	V: 0.01-0.10
	Grade C	$t \leq 200$	0.15	0.80-1.35	0.15-0.40	0.04	0.05	0.30-0.50	0.25-0.50	0.20-0.50	V: 0.01-0.10
	Grade K	$t \leq 200$	0.17	0.50-1.20	0.25-0.50	0.04	0.05	0.40-0.70	0.40	0.30-0.50	Nb: 0.005-0.05; Mo: 0.10
A 633	Grade A	$t \leq 100$	0.18	1.00-1.35	0.15-0.50	0.035	0.04	---	---	---	Nb: 0.05
	Grade C	$t \leq 100$	0.20	1.15-1.50	0.15-0.50	0.035	0.04	---	---	---	Nb: 0.01-0.05
	Grade D	$t \leq 40$	0.20	0.70-1.35	0.15-0.50	0.035	0.04	0.25	0.25	0.35	Mo: 0.08
		$40< t \leq 100$	0.20	1.00-1.60	0.15-0.50	0.035	0.04	0.25	0.25	0.35	Mo: 0.08
	Grade E	$t \leq 150$	0.22	1.15-1.50	0.15-0.50	0.035	0.04	---	---	---	V: 0.04-0.11; N: 0.03
A 656	Type 3	$t \leq 50$	0.18	1.65	0.60	0.025	0.035	---	---	---	V: 0.08; Nb: 0.008-0.10; N: 0.020
	Type 7	$t \leq 50$	0.18	1.65	0.60	0.025	0.035	---	---	---	V: 0.15; Nb: 0.10; N: 0.020
A 709	Grade 50	$t \leq 100$	0.23	1.35^{A}	0.40	0.04	0.05	---	---	0.20 min.^{C}	*D*
A 871	Type I	---	0.19	0.80-1.35	0.30-0.65	0.04	0.05	0.40-0.70	0.40	0.25-0.40	V: 0.02-0.10
	Type II	---	0.20	0.75-1.35	0.15-0.50	0.04	0.05	0.40-0.70	0.50	0.20-0.40	V: 0.01-0.10
	Type III	---	0.15	0.80-1.35	0.15-0.40	0.04	0.05	0.30-0.50	0.25-0.50	0.20-0.50	V: 0.01-0.10
	Type IV	---	0.17	0.50-1.20	0.25-0.50	0.04	0.05	0.40-0.70	0.40	0.30-0.50	Nb: 0.005-0.05; Mo: 0.10

Table 1.2. (Continued)

ASTM specification	Designation	Section thickness (mm)	Chemical compositions (wt.%, max.; unless otherwise specified)								
			C	Mn	Si	P	S	Cr	Ni	Cu	Others
A 1011	Grade 45 Class 1	---	0.22	1.35	---	0.04	0.04	0.15	0.20	0.20^{E}	V: 0.005 min.; Ti: 0.005 min.; Nb: 0.005 min.; Mo: 0.06
	Grade 45 Class 2	---	0.15	1.35	---	0.04	0.04	0.15	0.20	0.20^{E}	V: 0.005 min.; Ti: 0.005 min.; Nb: 0.005 min.; Mo: 0.06
	Grade 50 Class 1	---	0.23	1.35	---	0.04	0.04	0.15	0.20	0.20^{E}	V: 0.005 min.; Ti: 0.005 min.; Nb: 0.005 min.; Mo: 0.06
	Grade 50 Class 2	---	0.15	1.35	---	0.04	0.04	0.15	0.20	0.20^{E}	V: 0.005 min.; Ti: 0.005 min.; Nb: 0.005 min.; Mo: 0.06
	Grade 55 Class 1	---	0.25	1.35	---	0.04	0.04	0.15	0.20	0.20^{E}	V: 0.005 min.; Ti: 0.005 min.; Nb: 0.005 min.; Mo: 0.06
	Grade 55 Class 2	---	0.15	1.35	---	0.04	0.04	0.15	0.20	0.20^{E}	V: 0.005 min.; Ti: 0.005 min.; Nb: 0.005 min.; Mo: 0.06
	Grade 60 Class 1	---	0.26	1.50	---	0.04	0.04	0.15	0.20	0.20^{E}	V: 0.005 min.; Ti: 0.005 min.; Nb: 0.005 min.; Mo: 0.06
	Grade 60 Class 2	---	0.15	1.50	---	0.04	0.04	0.15	0.20	0.20^{E}	V: 0.005 min.; Ti: 0.005 min.; Nb: 0.005 min.; Mo: 0.06
	Grade 65 Class 1	---	0.26	1.50	---	0.04	0.04	0.15	0.20	0.20^{E}	V: 0.005 min.; Ti: 0.005 min.; Nb: 0.005 min.; Mo: 0.06
	Grade 65 Class 2	---	0.15	1.50	---	0.04	0.04	0.15	0.20	0.20^{E}	V: 0.005 min.; Ti: 0.005 min.; Nb: 0.005 min.; Mo: 0.06
	Grade 70 Class 1	---	0.26	1.65	---	0.04	0.04	0.15	0.20	0.20^{E}	V: 0.005 min.; Ti: 0.005 min.; Nb: 0.005 min.; Mo: 0.16
	Grade 70 Class 2	---	0.15	1.65	---	0.04	0.04	0.15	0.20	0.20^{E}	V: 0.005 min.; Ti: 0.005 min.; Nb: 0.005 min.; Mo: 0.16

[A] Values may vary, or minimum value may exist, depending on the product size and/or specified carbon content. The manganese to carbon ration shall not be less than 2 to 1. [B] Values may vary, or minimum value may exist, depending on the product size. [C] When specified. [D] Elements may be added as Type 1 (Nb: 0.005-0.05), Type 2 (V: 0.01-0.15), Type 3 (Nb: 0.005-0.05; V: 0.01-0.15; (Nb+V): 0.02-0.15), or Type 5 (Ti: 0.006-0.04; N: 0.003-0.015; V: 0.06 max). [E] When copper is specified, a minimum of 0.20% is required. When copper steel is not specified, the copper limit is the maximum requirement.

Table 1.3. Tensile properties of microalloyed steels

ASTM specification	Designation	Section thickness (mm)	YS, min.		TS, min.		EL (%), min.	
			Mpa	ksi	Mpa	ksi	GL=200 mm	GL=50 mm
A 242	Type 1	$t \le 20$	345	50	480	70	18	21
		$20 < t \le 40$	315	46	460	67	18	21
		$40 < t \le 100$	290	42	435	63	18	21
A 572	Grade 42	---	290	42	415	60	20	24
	Grade 50	---	345	50	450	65	18	21
	Grade 55	---	380	55	485	70	17	20
	Grade 60	---	415	60	520	75	16	18
	Grade 65	---	450	65	550	80	15	17
A 588	---	$t \le 100$	345	50	485	70	18	21
	---	$100 < t \le 125$	315	46	460	67	---	21
	---	$125 < t \le 200$	290	42	435	63	---	21
A 633	Grade A	$t \le 65$	290	42	430-570	63-83	18	23
		$65 < t \le 100$	290	42	430-570	63-83	18	23
		$100 < t \le 150$	---	---	---	---	18	23
	Grade C,	$t \le 65$	345	50	485-620	70-90	18	23
	Grade D	$65 < t \le 100$	315	46	450-590	65-85	18	23
		$100 < t \le 150$	---	---	---	---	18	23
	Grade E	$t \le 65$	415	60	550-690	80-100	18	23
		$65 < t \le 100$	415	60	550-690	80-100	18	23
		$100 < t \le 150$	380	55	515-655	75-95	18	23
A 656	Grade 50	---	345	50	415	60	20	23
	Grade 60	---	415	60	485	70	17	20
	Grade 70	---	485	70	550	80	14	17
	Grade 80	---	550	80	620	90	12	15
A 709	Grade 50	$t \le 100$	345	50	450	65	18	21
A 871	Grade 60	---	415	60	520	75	16	18
	Grade 65	---	450	65	550	80	15	17

Table 1.3. (Continued)

ASTM specification	Designation	Section thickness (mm)	YS, Min.		TS, Min.		EL (%), Min.	
			Mpa	ksi	Mpa	ksi	GL=200 mm	GL=50 mm
A 1011	Grade 45 Class 1	t ≤ 2.5	310	45	410	60	---	23
		t > 2.5	310	45	410	60	---	25
	Grade 45 Class 2	t ≤ 2.5	310	45	380	55	---	23
		t > 2.5	310	45	380	55	---	25
	Grade 50 Class 1	t ≤ 2.5	340	50	450	65	---	20
		t > 2.5	340	50	450	65	---	22
	Grade 50 Class 2	t ≤ 2.5	340	50	410	60	---	20
		t > 2.5	340	50	410	60	---	22
	Grade 55 Class 1	t ≤ 2.5	380	55	480	70	---	18
		t > 2.5	380	55	480	70	---	20
	Grade 55 Class 2	t ≤ 2.5	380	55	450	65	---	18
		t > 2.5	380	55	450	65	---	20
	Grade 60 Class 1	t ≤ 2.5	410	60	520	75	---	16
		t > 2.5	410	60	520	75	---	18
	Grade 60 Class 2	t ≤ 2.5	410	60	480	70	---	16
		t > 2.5	410	60	480	70	---	18
	Grade 65 Class 1	t ≤ 2.5	450	65	550	80	---	14
		t > 2.5	450	65	550	80	---	16
	Grade 65 Class 2	t ≤ 2.5	450	65	520	75	---	14
		t > 2.5	450	65	520	75	---	16
	Grade 70 Class 1	t ≤ 2.5	480	70	585	85	---	12
		t > 2.5	480	70	585	85	---	14
	Grade 70 Class 2	t ≤ 2.5	480	70	550	80	---	12
		t > 2.5	480	70	550	80	---	14

Note: where "---" appears in this table, there is no requirement. "GL"=Gage Length, "YS"=Yield Strength, "TS"=Tensile Strength, and "EL"=Elongation.

Table 1.4. Chemical compositions of microalloyed steels studied in this book (wt.%)

Steels	C	Mn	Cr	Ni	Cu	Si	Al	V	Nb	Ti	P	S	N	W
0W	0.171	1.22	0.103	0.025	0.023	0.469	0.019	0.015	0.023	0.019	0.013	0.008	0.007	**0**
0.1W	0.170	1.20	0.103	0.024	0.024	0.471	0.024	0.015	0.023	0.020	0.014	0.009	0.008	**0.09**
0.5W	0.167	1.20	0.113	0.021	0.022	0.497	0.024	0.005	0.018	0.023	0.008	0.008	0.005	**0.47**
1W	0.171	1.20	0.112	0.024	0.022	0.499	0.021	0.016	0.020	0.020	0.008	0.008	0.006	**0.99**
2W	0.170	1.22	0.092	0.030	0.021	0.500	0.014	0.012	0.021	0.019	0.009	0.008	0.005	**2.04**

The microstructural evolution of microalloyed steels including grain size, phase component, pearlite interlamellar spacing, allotriomorphic ferrite and acicular ferrite after tungsten addition is then considered, followed by a chapter that describes the effect of tungsten on mechanical properties of microalloyed steels by means of hardness, impact toughness and room/low-temperature tensile properties tests. Mechanical properties depend directly on microstructure, and therefore, microstructure-property relationships are analysed in detail. Next, heat treatment technologies of tungsten-containing microalloyed steels are introduced with the purpose of achieving the best combination of excellent strength and toughness. With the applications of microalloyed steels in offshore structures and hydrogen environment, the corrosion resistance and hydrogen embrittlement become two very important criteria for assessing the steel's performance. The effects of tungsten on the corrosion resistance and hydrogen embrittlement of microalloyed steels are therefore discussed in the following two chapters. Finally, conclusion remarks and the problems faced at present and the possibility for future development of tungsten alloying in microalloyed steels are stated.

REFERENCES

[1] D. R. Lide (Ed.), *CRC Handbook of Chemistry and Physics*, 75th ed., CRC Press, Boca Raton, 1994.

[2] E. Lassner, W.D. Schubert, *Tungsten: Properties, Chemistry, Technology of the Elements, Alloys, and Chemical Compounds*, Kluwer Academic / Plenum Publishers, New York, 1999.

[3] J. Emsley, *The Elements*, 2nd ed., Oxford University Press, New York, 1991.

[4] P.M. Morse, Q.D. Shelby, D.Y. Kim, G.S. Girolami, Ethylene Complexes of the Early Transition Metals: Crystal Structures of $[HfEt_4(C_2H_4)^{2-}]$ and the Negative-Oxidation-State Species $[TaHEt(C_2H_4)_3^{3-}]$ and $[WH(C_2H_4)_4^{3-}]$, *Organometallics*, 27(5) (2008) 984-993.

[5] D. Shan, Y. Xie, G. Ren, Z. Yang, Inhibitory effect of tungsten carbide nanoparticles on voltage-gated potassium currents of hippocampal CA1 neurons, *Toxicology Letters*, 209(2) (2012) 129-135.

[6] K. Schultz, Ken Schultz's Essentials of Fishing: *The Only Guide You Need to Catch Freshwater and Saltwater Fish*, John Wiley & Sons, Inc., Hoboken, 2009.

[7] Wolf-Dieter Schubert, Erik Lassner, Herbert Danninger, Tungsten in steel. *ITIA Newsletter*, December 2008, pp. 2-11.

[8] ASTM A600 - 92a, Standard Specification for Tool Steel High Speed, *ASTM International*, 2010.

[9] G.A. Roberts, R. Kennedy, *Tool Steels*, ASM International, 1998.

[10] C.R. Sohar, *Lifetime Controlling Defects in Tool Steels*, Springer, 2011.

[11] Z.F. Hu, Heat-Resistant Steels, Microstructure Evolution and Life Assessment in Power Plants, in: M. Rasul (Ed.), *Thermal Power Plants, InTech,* 2012, pp. 195-226.

[12] W. Sha, Steels: *From Materials Science to Structural Engineering*, Springer, 2013.

[13] J.S. Park, S.J. Kim, C.S. Lee, Effect of W addition on the low cycle fatigue behavior of high Cr ferritic steels, *Materials Science and Engineering A*, 298 (2001) 127-136.

[14] J.R. Davis (Ed.), *Heat-Resistant Materials*, ASM International, 1997.

[15] G. Krauss, *Steels: Processing, Structure, and Performance*, ASM International, 2005.

[16] A. Irhzo, Y. Segui, Z. Bui, F. Dabosi, The role of alloyed tungsten on the conductivity of stainless steel passive layers, *Corrosion Science*, 26 (1986) 769-780.

[17] K. Laha, S. Saroja, A. Moitra, R. Sandhya, M.D. Mathew, T. Jayakumar, E. Rajendra Kumar, Development of India-specific RAFM steel through optimization of tungsten and tantalum contents for better combination of impact, tensile, low cycle fatigue and creep properties, *Journal of Nuclear Materials*, 439 (2013) 41-50.

[18] P. Dou, A. Kimura, R. Kasada, T. Okuda, M. Inoue, S. Ukai, S. Ohnuki, T. Fujisawa, F. Abe, Effects of titanium concentration and tungsten addition on the nano-mesoscopic structure of high-Cr oxide dispersion strengthened (ODS) ferritic steels, *Journal of Nuclear Materials*, 442 (2013) S95-S100.

[19] Z. Grzesik, G. Smola, K. Adamaszek, Z. Jurasz, S. Mrowec, High temperature corrosion of valve steels in combustion gases of petrol containing ethanol addition, *Corrosion Science*, 77 (2013) 369-374.

[20] Y. Zhang, Y. Sun, S. Guan, X. Deng, X. Yan, Effect of titanium and tungsten on the structure and properties of heat-abrasion resistant steel, *Materials Science and Engineering A*, 478 (2008) 214-220.

[21] W.J. Nam, C.S. Lee, D.Y. Ban, Improvement of sag resistance by the addition of tungsten in Si-Cr-Mo-V steels, *Scripta Materialia*, 36 (1997) 1315-1320.

[22] J.I. Suk, C.N. Park, S.H. Hong, Y.G. Kim, Development and properties of tungsten-bearing stainless maraging steels, *Materials Science and Engineering A*, 138 (1991) 267-273.

[23] N.H. Heo, H.C. Lee, Effect of tungsten addition on the ductile-brittle-ductile transition in Fe-8Mn-7Ni-W maraging steels, *Scripta Metallurgica et Materialia*, 33 (1995) 2031-2035.

[24] H.R. Yang, K.B. Lee, H. Kwon, Effects of austenitizing treatments and inclusion on secondary hardening and fracture behavior for high Co-Ni steels containing W, *Materials Science and Engineering A*, 265 (1999) 179-187.

[25] J.R. Davis, *Alloying: Understanding the Basics*, ASM International, 2001.

[26] D.W. Oh, *The Influence of Boron and Titanium on Low Carbon Microalloyed Steel Weld Metals*, Colorado School of Mines, 1987.

[27] H. Zhang, C. Liu, G. Pang, Research of low carbon Nb-Ti-B microalloyed high strength hot strip steels with yield strength ≥700 MPa, in: Y. Weng, H. Dong, Y. Gan (Eds.), *Advanced Steels: The Recent Scenario in Steel Science and Technology*, Springer, 2011, pp. 317-332.

[28] D.B. Park, M.Y. Huh, J.H. Shim, J.Y. Suh, K.H. Lee, W.S. Jung, Strengthening mechanism of hot rolled Ti and Nb microalloyed HSLA steels containing Mo and W with various coiling temperature, *Materials Science and Engineering A*, 560 (2013) 528-534.

[29] R. Lagneborg, T. Siwecki, S. Zajac, B. Hutchinson, The role of vanadium in microalloyed steels, *Scandinavian Journal of Metallurgy*, 28(5) (1999) 186-241.

[30] T.N. Baker, Process, microstructure and properties of vanadium microalloyed steels, *Materials Science and Technology*, 25(9) (2009) 1083-1107.

[31] T.H. Xi, X. Chen, P.H. Li, Z.X. Yuan, Effects of zirconium on microstructures and mechanical properties of microalloyed steels, *Acta Metallurgica Sinica (English Letters)*, 19(5) (2006) 319-327.

[32] A.R.B. Maia, C.R. Guinâncio, R.L. Germano, P.R. Rion, I.S. Bott, Use of zirconium in microalloyed steels, *Advanced Materials Research*, 15-17 (2006) 834-839.

[33] R.D.K. Misra, Z. Jia, R. O'Malley, S.J. Jansto, Precipitation behavior during thin slab thermomechanical processing and isothermal aging of copper-bearing niobium-microalloyed high strength structural steels: The effect on mechanical properties, *Materials Science and Engineering A*, 528 (2011) 8772-8780.

[34] M. Opiela, M. Kamińska, Influence of the rare-earth elements on the morphology of non-metallic inclusions in microalloyed steels, *Journal of Achievements in Materials and Manufacturing Engineering*, 47(2) (2011) 149-156.

[35] H.Y. Wang, H.P. Ren, H. Yang, S.D. Wang, D.X. Li, Effect of rare earth addition on dynamic recrystallization and precipitation behavior of a niobium-containing microalloy steel, *Materials Science Forum*, 762 (2013) 189-193.

[36] ASTM A242/A242M, Standard Specification for High-Strength Low-Alloy Structural Steel, *ASTM International*, 2004.

[37] ASTM A572/A572M, Standard Specification for High-Strength Low-Alloy Columbium-Vanadium Structural Steel, *ASTM International*, 2007.

[38] ASTM A588/A588M, Standard Specification for High-Strength Low-Alloy Structural Steel, up to 50ksi [345 MPa] Minimum Yield Point, with Atmospheric Corrosion Resistance, *ASTM International*, 2005.

[39] ASTM A633/A633M, Standard Specification for Normalized High-Strength Low-Alloy Structural Steel Plates, *ASTM International*, 2001.

[40] ASTM A656/A656M, Standard Specification for Hot-Rolled Structural Steel, High-Strength Low-Alloy Plate with Improved Formability, *ASTM International*, 2003.

[41] ASTM A709/A709M, Standard Specification for Structural Steel for Bridges, *ASTM International*, 2009.

[42] ASTM A871/A871M, Standard Specification for High-Strength Low-Alloy Structural Steel Plate with Atmospheric Corrosion Resistance, *ASTM International*, 2003.

[43] ASTM A1011/A1011M, Standard Specification for Steel, Sheet and Strip, Hot-Rolled, Carbon, Structural, High-Strength Low-Alloy, High-Strength Low-Alloy with Improved Formability, and Ultra-High Strength, *ASTM International*, 2007.

Chapter 2

EFFECT OF TUNGSTEN ON PHASE TRANSFORMATION

ABSTRACT

The desired mechanical properties of microalloyed steels are developed by manipulation of different microstructural combinations through suitable reheating and cooling cycles. A knowledge base of solid-state phase transformations that occur in microalloyed steels is required in order to understand well the relationship between microstructure and mechanical properties. This chapter describes the iron-carbon equilibrium diagram and the phase transformations occurred in microalloyed steels. The isothermal and continuous cooling transformation diagrams are introduced with purpose of providing an in-depth understanding of the phase transformation mechanism. Continuous cooling transformation diagrams of microalloyed steels with different tungsten contents have been created and presented. The phase transformation characteristics of these steels at different cooling rates are analysed in detail in order to provide necessary information for application to practical processes. The effects of tungsten on critical cooling rate for phase transformation and microstructural components, phase transformation temperature, transformed microstructure and hardness after phase transformation are detailed. The research presented in this chapter has immediate practical applications, and the outcomes can be applied in both heat treatment and hot working processes of microalloyed steels in steel manufacturing industries.

2.1. Iron-Carbon Equilibrium Diagram

The iron-carbon (Fe-C) equilibrium diagram shown in Figure 2.1 is the basis for understanding the microstructures in microalloyed steels. In the figure, α-iron (ferrite) has a BCC crystal structure, and is stable from all temperatures up to 912°C when it transforms into the γ-iron (austenite) which has a face-centred cubic (FCC) crystal structure. γ-iron exists between 912 and 1394°C. δ-iron has also a BCC crystal structure, and exists stably from 1394°C to iron melting point of 1538°C. The amount of carbon determines whether a material is steel or iron. Iron based materials with carbon content below than 2.11% are classified as steels. Iron alloys containing more than 2.11% carbon are called cast irons because of their dominant iron content, low melting points and good castability. The diagram describes the Fe-C system of alloys containing up to 6.67% of carbon which corresponds to the fixed composition of the iron carbide Fe_3C (known as cementite). Cementite has an orthorhombic crystal structure and is a hard and brittle phase. Cementite phase is metastable with respect to iron and graphite equilibrium [1]. Although graphite occurs extensively in cast irons, it is usually difficult to obtain this equilibrium phase in steels. Hence, the metastable equilibrium between iron and cementite should be considered for all discussions on phase transformations and heat treatment of steels due to its relevance to the behaviour of most steels in practice.

There are several critical temperatures in the Fe-C equilibrium diagram which are important from both the basic and practical points of view:

A_1: The temperature at which eutectoid reaction occurs. Below this temperature austenite does not exist.
A_3: The temperature below which ferrite starts to form as a result of ejection from austenite in the hypo-eutectoid steels.
A_{cm}: The temperature below which cementite starts to form as a result of ejection from austenite in the hyper-eutectoid steels.
A_2: The Curie temperature below which α-iron is paramagnetic.
A_4: The temperature below which δ-ferrite transforms to γ-iron.

The A_1, A_3 and A_4 temperatures can be easily detected by thermal analysis or dilatometer tests during heating or cooling cycles. Due to the presence of hysteresis in practical measurements, subscript letters "c" and "r" are generally used to identify critical temperatures under different heating or cooling conditions, i.e., A_c (the "c" from the French word *chauffage*) for heating, and A_r (the "r" from the French word *refroidissement*) for cooling. A_e is often used to represent equilibrium. For example, A_{c1} and A_{c3} are respectively the ferrite-to-austenite transformation starting and finishing temperatures during heating, and A_{r3} and A_{r1} are the austenite-to-ferrite transformation starting and finishing temperatures during cooling. A_c and A_r values are sensitive to the heating and cooling rates, and as well as the presence of alloying elements.

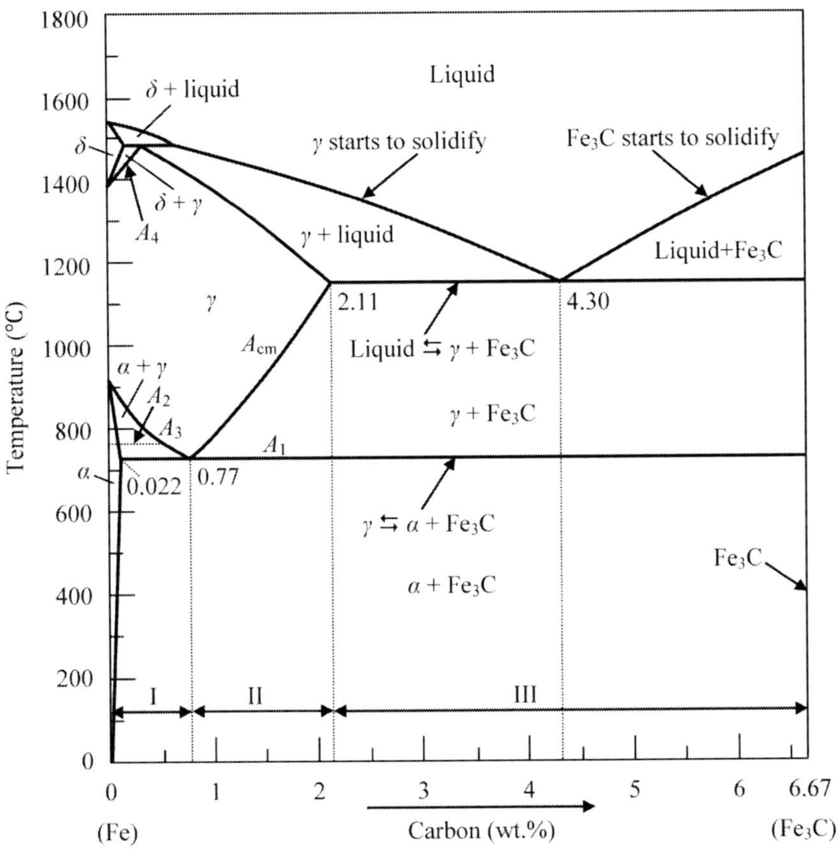

Figure 2.1. Fe-C equilibrium phase diagram (schematic): I=hypo-eutectoid steels, II=hyper-eutectoid steels, and III=cast irons.

Carbon dissolves interstitially in iron. The solubility of carbon in γ-iron is bigger than that in α-iron. The maximum solubility of carbon in γ-iron is 2.11% at 1148°C. The great difference in carbon solubility in γ-iron and α-iron leads to the rejection of carbon as iron carbide at the boundaries of the γ-iron field. On cooling steels containing less than 0.77% carbon slowly, hypo-eutectoid ferrite forms with enrichment of the residual austenite in carbon until at 727°C. Now the remaining austenite contains 0.77% carbon, and eutectoid reaction will occur during the subsequent slow cooling to form the microstructure of pearlite, a lamellar mixture of alternating layers of ferrite and cementite. For the hyper-eutectoid steels containing 0.77-2.11% carbon, cementite first forms progressively depleting the austenite in carbon when cooling slowly until at 727°C, the austenite contains 0.77% carbon and will form pearlite.

It should be noted that the Fe-C equilibrium diagram can be considered only as a guide because all microalloyed steels contain various alloying elements that may modify the positions of phase boundaries. Also, the Fe-C equilibrium is further limited due to the fact that some heat treatments are specifically designed to produce non-equilibrium structures. Nevertheless, the Fe-C diagram provides us a valuable scientific basis for the development of heat treatment and other processing technologies of microalloyed steels in their immense variety.

2.2. Phase Transformations in Microalloyed Steels

Several phase transformations including ferritic transformation, pearlitic transformation, bainitic transformation and martensitic transformation may be involved during heat treatment of microalloyed steels. Austenite is the parent phase of all the microstructures including pearlite, bainite, martensite and various ferritic microstructures. An in-depth understanding of the transformations from austenite to ferrite, pearlite, banite and martensite is of significance for the obtaining of optimal final microstructures with excellent properties. This section will therefore focus on the phase transformations that will occur in microalloyed steels when cooling is performed from austenite region.

2.2.1. Austenite-to-Ferrite Transformation

The austenite-to-ferrite transformation can be classified into the following five distinct ferrite morphologies based on Dubé classification [2]. The ferrite morphologies are schematically illustrated in Figure 2.2.

(i) *Grain boundary allotriomorphs.* They are crystals which nucleate at the austenite grain boundaries and then grow along these boundaries as well as perpendicular to them. The allotriomorph is in contact with at least two of the austenite grains and will have a random orientation with one of them, but an orientation which is more coherent with the other. The growth rates of allotriomorphs along the grain boundaries are generally much faster relative to those normal to them.

(ii) *Widmanstätten ferrite plates or laths.* These plates grow along well-defined planes into the interior of austenite grains from the grain boundaries or from the existing grain boundary allotriomorphs but do not grow across the austenite grain boundaries.

(iii) *Intragranular idiomorphs.* These are roughly equiaxed crystals that nucleate inside the austenite grains. An idiomorph forms without contact with the austenite grain surfaces and has a shape which some shows crystallographic facets.

(iv) *Intragranular Widmanstätten plates.* These plates are crystals similar to those growing from grain boundaries. Differently, they nucleate entirely within the austenite grains.

(v) *Massive ferrite structure.* This structure is a result of the impingement of crystals from other morphologies, and of grain growth within the aggregate, to form bulky, roughly equiaxed polycrystalline masses in the microstructure.

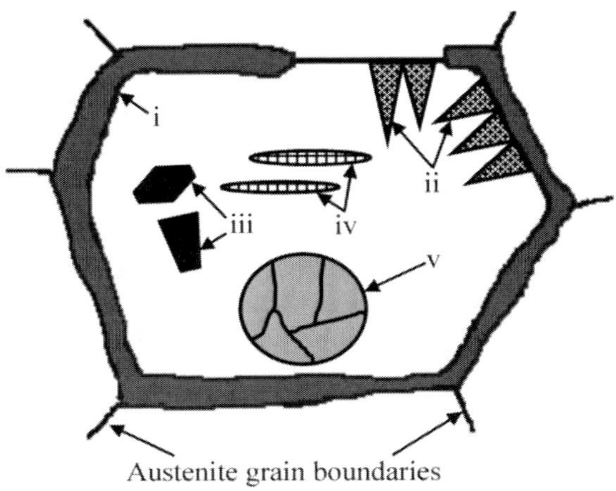

Figure 2.2. Dubé classification of ferrite morphologies: i= allotriomorphic ferrite, ii=Widmanstätten ferrite, iii=intragranular idiomorphic ferrite, iv=intragranular Widmanstätten ferrite, and v=massive ferrite structure.

Microalloyed steels belong to hypo-eutectoid steels. Consider a hypo-eutectoid steel, after austenitisation, is allowed to transform to ferrite at various temperatures below A_3, as schematically shown in Figure 2.3. At very slow undercooling below A_3, ferrite nucleates at austenite grain boundaries and grows along the grain boundary contours to form grain boundary allotriomorphs. Allotriomorphic ferrite is the first phase to form on cooling the austenite grains below A_3. The allotriomorphic ferrite has a reproducible orientation relationship with one austenite grain (γ_1) which is close to the Kurdjumov-Sachs orientation type. But they should normally have a random orientation relationship because they also grow into the adjacent austenite grain (γ_2). The tendency for the Widmanstätten ferrite formation increases as undercooling increases, and the larger the undercooling, the finer the Widmanstätten ferrite will be formed. As the reason for the transition from grain boundary allotriomorphic ferrite to Widmanstätten ferrite, Aaronson et al. [3] suggested that the relative rates at which semi-coherent and incoherent interfaces can migrate and vary with undercooling. At small undercooling, both semi-coherent and incoherent interfaces can migrate at similar rates, while only incoherent interfaces can make full use of the increased driving force at relatively large undercooling. Widmanstätten ferrite plates have a definite crystallographic orientation relationship with the austenite grains into which they grow as [2]:

$$(110)_\alpha \parallel (111)_\gamma$$
$$[1\bar{1}1]_\alpha \parallel [1\bar{1}0]_\gamma$$

The above orientation relationship indicates that the close packed planes of crystal lattices of α and γ and the closed packed directions in these planes are parallel to each other. This relationship exists along the broad faces of the Widmanstätten ferrite plates, and the interface between α and γ along these faces is semi-coherent. Generally, such interface has very low mobility.

In fact, ferrite can also nucleate within austenite grains. Non-metallic inclusions and dislocations are thought to be the suitable heterogeneous nucleation sites for such ferrite precipitation. This kind of ferrite generally has equiaxed shape at low undercooling, and plate-like at higher undercooling. The nucleation rate within austenite grains is generally less than that at austenite grain boundaries, and the precipitation of intragranular ferrite depends on the austenite grain size. A large austenite grain size encourages the intragranular ferrite nucleation because in a coarse-grained austenite it takes a longer time for the carbon rejected from the ferrite that forms at grain boundaries to reach the centre of the grain, and thereby increasing the time for ferrite nucleation at the intragranular sites.

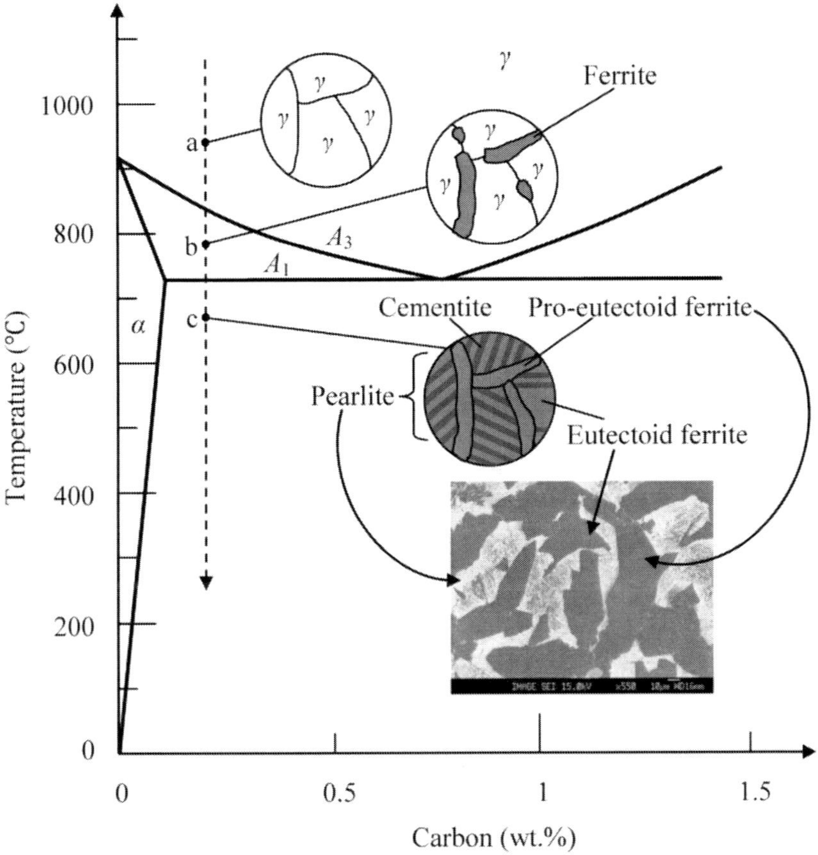

Figure 2.3. Schematic and real microstructures of a hypo-eutectoid steel cooled slowly from austenite.

2.2.2. Austenite-to-Pearlite Transformation

When the temperature is decreased below A_1, austenite becomes simultaneously supersaturated with both ferrite and cementite, and a eutectoid reaction will occur:

$\gamma \rightarrow \alpha + Fe_3C$

The resultant microstructure is composed of lamellar cementite embedded alternately in ferrite what is known as pearlite, as schematically shown in Figure 2.3.

The scanning electron microscope (SEM) image embedded in the figure illustrates the real microstructure of a microalloyed steel after austenite-to-pearlite transformation.

Like ferritic transformation, the pearlitic transformation also occurs by a process of nucleation and growth. Pearlite nucleates heterogeneously at the austenite grain boundaries with a roughly constant radial velocity into the surrounding austenite grains.

The nucleation of pearlite requires the establishment of cooperative growth of the two phases (ferrite and cementite). The first stage in the formation of pearlite is the nucleation of either ferrite or cementite at austenite grain boundaries.

The nuclei that form first depend on the grain boundary structure and composition, and will determine the nucleation rate. When undercooling below A_1 is low, the pearlite nodules can grow at a constant rate in all directions as hemispheres or spheres without interfering with each other due to the low nucleation rate and the presence of only small quantity of nucleated pearlite nodules.

At relatively large undercooling, where the nucleation rate is high, pearlite nodules nucleated at austenite grain boundaries soon impinge on each other, and all grain boundaries will be quickly covered with nodules which grow together forming layers of pearlite with an essentially planar front into the austenite grains.

Figure 2.4 shows a real SEM image of pearlite observed in a microalloyed steel. In the figure, the white laths are cementite lamellae, and the grey region indicates ferrite phase.

The growth of pearlite requires redistribution of carbon. The carbon concentration in the ferrite phase in pearlite is negligible, whereas the carbon concentration of the cementite is about 6.67%.

During the growth of pearlite, carbon is redistributed either by the diffusion of carbon in austenite matrix ahead of the austenite/pearlite interface or by phase boundary diffusion through the austenite/pearlite itself.

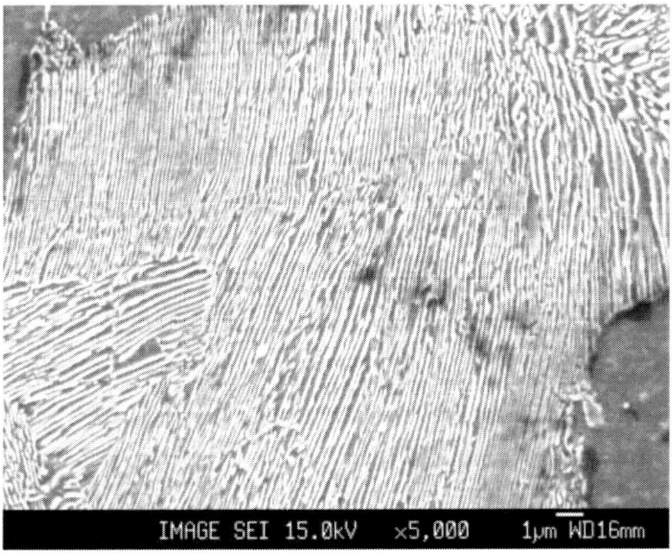

Figure 2.4. SEM image of pearlite observed in a microalloyed steel.

2.2.3. Austenite-to-Bainite Transformation

Bainite reaction takes place at temperatures from 250 to 550°C in Fe-C alloys of all compositions [1]. Bainite is a mixture of ferrite and carbide, but it is quite distinct from pearlite. The pearlitic transformation involves the redistribution of carbon followed by a structure change, whereas bainitic transformation involves a structure change followed by the redistribution of carbon, which precipitates as a carbide. The morphology of bainite depends mainly on the temperature at which it forms. Bainite is generally classified into upper and lower bainite based on the microstructural features [4].

Upper bainite, formed at high temperatures (400-550°C), consists of fine plates of ferrite with carbide (cementite) precipitates between theses plates. For the upper bainite reaction, bainitic ferrite which has very low carbon solubility will be first formed.

The growth of the ferrite enriches the remaining austenite in carbon, and eventually cementite precipitates from the residual austenite layers in between the ferrite sub-units. The amount of cementite depends on the carbon concentration of the alloy, i.e., high carbon concentration leads to microstructures in which the ferrite plates are separated by continuous layers of cementite, whereas small, discrete particles of cementite will form when the carbon concentration is low. The ferrite plates grow into the austenite grains in a similar way to Widmanstätten ferrite plates. The growth of upper bainite is essentially controlled by the growth of ferrite plates, and cementite forms in the carbon-enriched regions adjacent to these plates. A number of ferrite plates together form a packet.

Within a packet, the plates are separated by low angle grain boundaries, whereas the boundaries between different packets are high angle boundaries. At the highest temperatures where pearlite and bainite grow competitively, it is difficult to distinguish the pearlite colonies from the upper bainite because both appear as alternate layers of cementite in ferrite. One means of distinguishing pearlite from bainte is by their crystallography. In pearlite, the cementite and ferrite have no specific orientation relationship with austenite grain they are growing. In the case of bainite, however, the cementite and ferrite do have an orientation relationship with austenite grain they are growing. The cementite in bainite is rigorously crystallographic in form, and it has a Pitsch orientation relationship with austenite from which it precipitates [1]:

$$[100]_{\text{cementite}} \parallel [5\bar{5}4]_{\gamma}$$
$$[010]_{\text{cementite}} \parallel [\bar{1}\,\bar{1}\,0]_{\gamma}$$
$$[001]_{\text{cementite}} \parallel [\bar{2}25]_{\gamma}$$

At sufficiently low temperatures (250-400°C), lower bainite reaction will occur. Lower bainite has a very similar microstructure and crystallographic features to those of upper bainite. The microstructure of lower bainite is composed of acicular ferrite, with fine carbide precipitation occurring essentially within these laths. The carbide precipitates have a Bagaryatski crystallographic orientation relationship with the ferrite laths in which they form [1]:

$$[100]_{\text{cementite}} \parallel [111]_{\alpha}$$
$$[010]_{\text{cementite}} \parallel [\bar{1}2\bar{1}]_{\alpha}$$
$$[001]_{\text{cementite}} \parallel [\bar{1}01]_{\alpha}$$

The carbides in lower bainite may not always be cementite. The carbides that precipitated inside the ferritic laths could be metastable transition carbides such as hexagonal ε-carbide rather than cementite depending on the chemical compositions and transformation temperatures. The carbides precipitated in lower bainite are extremely fine, and they distribute more uniformly relative to that in upper bainite, where cementite precipitates as thin films between carbide-free ferritic laths.

Lower bainite is usually observed to be much tougher than upper bainite, in spite of the fact that is tends to be stronger. The coarse cementite laths in upper bainite favour the nucleation of cleavage cracks and voids.

2.2.4. Austenite-to-Martensite Transformation

When a microalloyed steel is rapidly cooled, such as quenched, from austenite region to room temperature at a cooling rate sufficient to supress its decomposition by diffusional processes to ferrite, pearlite and/or bainite, a hard and brittle phase called martensite will be formed. The austenite-to-martensite transformation is diffusionless, and therefore martensite forms without any interchange in the position of neighbouring atoms. To verify the diffusionless nature of martensitic transformation, the chemical composition of martensite can be measured and should be identical to that of the parent austenite. There is a specific temperature, M_s, at which the martensitic transformation starts. The fraction of the transformed martensite is generally a function of temperature and it increases continuously from zero at M_s to completion at martensite finishing temperature M_f. The M_f temperature corresponds to that temperature below which further cooling does not increase the amount of martensite. In practice, however, the M_f may not correspond to 100% martensite, and some retained austenite can be left even below M_f. Martensitic transformation takes place at a very rapid speed, and the time taken for the transformation to occur at a given temperature is generally negligible. If the cooling stops at a temperature between M_s and M_f, the transformation will be halted and it may proceed only on further cooling.

Martensite in microalloyed steels has a body-centred tetragonal (BCT) crystal structure by quenching the austenite phase which may be regarded as a ferrite lattice with one of the cubic axes elongated. The crystal structure of martensite is relatively closer to that of ferrite than austenite. The c/a ratio in the BCT structure varies with the carbon concentration of martensite, and equals to one is the same as the ferrite BCC structure and the c/a ratio varies from ~1.0 for very low carbon steels to ~1.1 for high carbon steels. The symbol α' is therefore often used to denote martensite since its crystal structure may be thought to be derived from the crystal structure of ferrite (α) phase.

The martensitic transformation also starts with nucleation. A necessary condition for the occurrence of the austenite-to-martensite transformation is that the free energy of martensite be lower than that of austenite. The driving force for the nucleation of martensite at the M_s temperature can be given by:

$$\Delta G^{\gamma\to\alpha'} = \Delta H^{\gamma\to\alpha'} \frac{(T_0 - M_s)}{T_0} \tag{2.1}$$

where $\Delta G^{\gamma\to\alpha'}$ is the free energy change during the martensitic transformation at the M_s temperature, ΔH is the enthalpy variation during the martensitic transformation, T_0 is the temperature at which the free energy difference between austenite and martensite of the same composition is zero. The austenite-to-martensite transformation cannot take place until the steel is cooled to a particular temperature M_s below T_0. The larger the difference between the two crystal structures, the greater the degree of undercooling will be, because it is more difficult for the change to occur when greatly differing structures are involved. In the case of microalloyed steels, the difference between the two structures is rather large, and the difference between M_s below T_0 would be very large. Such great difference in crystal structure will cause a very low M_s temperature below the extended A_3 line shown in Figure 2.1. In microalloyed steels, the martensite formed in an austenite has a definite Kurdjumov-Sachs (K-S) crystallographic orientation relationship with those of the untransformed part of austenite as [5]:

$$(111)\gamma \parallel (011)_{\alpha'}$$
$$[\bar{1}01]\gamma \parallel [\bar{1}\,\bar{1}\,1]_{\alpha'}$$

Martensite exists microstructurally in two principal morphologies i.e., lath martensite and acicular martensite. In microalloyed steels, lath martensite is the common morphology.

Figure 2.5. OM structure of martensite observed in a microalloyed steel that was water-quenched from 1100°C.

Figure 2.5 shows an optical microscope (OM) structure of martensite observed in a microalloyed steel that was water-quenched from 1100°C. The crystals in lath martensite are in the form of thin plates extending in one direction. A number of these crystals which are essentially parallel to each other together form dense packets. Within a packet, the adjacent crystals are separated by low angle grain boundaries. The width of crystals within a packet is roughly the same and it varies from ~0.1 to a few μm. The lath martensite crystals generally contain high density of dislocations with the order of 10^{11} - 10^{12} cm^{-2} [2].

2.3. Isothermal and Continuous Cooling Transformation Diagrams

2.3.1. Isothermal Transformation Diagram

Isothermal transformation diagram, also known as time-temperature transformation (TTT) diagram, of steel is a representation of the overall kinetics of isothermal transformation of austenite to low temperature products including ferrite, pearlite, bainite and martensite at different temperatures.

TTT diagram plots time on the *X*-axis (usually on a logarithmic scale) versus temperature on the *Y*-axis. A typical TTT diagram for a hypo-eutectoid steel is schematically shown in Figure 2.6. This kind of diagram, made up from the results of a series of isothermal decomposition experiments, shows the time required for transformation to start and to finish as a function of temperature. Since it is difficult to measure experimentally the time corresponding to the starting of the transformation, the starting curve usually corresponds to 1% austenite decomposition, and similarly, the end curve may represent 99% austenite decomposition. A TTT diagram allows one to determine the end products of homogeneous austenite transformation upon its cooling by holding the steel at a fixed temperature below the A_1 critical temperature.

In Figure 2.6, the nose of the TTT diagram indicates the minimum time (t_0) required for the start of austenite decomposition occurs at a temperature where the rates of nucleation and growth are normally the highest. The nose of a TTT diagram is an important parameter because it may be used to determine the minimum cooling rate to obtain 100% martensitic transformation. When the temperature is slightly lower than the critical temperature, A_1, the transformation rate is very slow. It is because that the necessary surface and strain energy contribute to the nucleation for the phase change (e.g., austenite-to-pearlite transformation) is not sufficient enough even though the atomic mobility must be high in such temperature range.

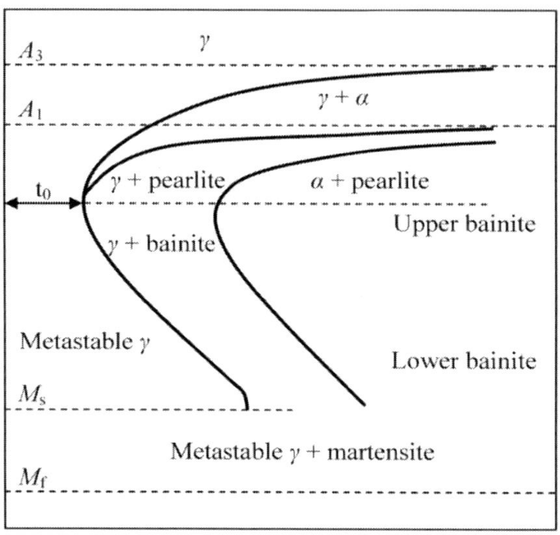

Figure 2.6. Schematic TTT diagram for a hypo-eutectoid steel.

The transformation rate increases as the temperature approaches the nose of the TTT starting curve. When bainite is formed as the temperature drops below the nose of the TTT diagram, the transformation rate becomes slow again. This is because that bainitic transformation depends on diffusion; however the atomic migration at theses temperatures is slow. The austenite-to-martensite transformation is diffusionless, and it does not depend on the speed of migration of atoms. The austenite only starts transforming to martensite when the temperature falls below a critical temperature M_s. This explains why the martensitic transformation speeds up again even though atomic mobility in this temperature range is very low. M_s in the diagram is identified as the highest temperature for which martensite can be detected in the microstructure. Below M_s, the volume fraction of martensite transformed from austenite increases to unity at M_f. For experimental convenience however, the M_s is usually determined at 0.01 volume fraction of martensite, and M_f at 0.99 volume fraction. The values of M_s and M_f can be predicted from the composition of the austenite by empirical relationships that have been formulated [6-10]. For example, according to Andrews [6]:

$$\begin{aligned} M_s &= 539 - 423C - 30.4Mn - 17.7Ni - 12.1Cr - 11.0Si - 7.0Mo \\ M_s &= 512 - 453C - 16.9Ni - 9.5Mo + 217C^2 - 71.5C \cdot Mn + 15Cr - 67.6C \cdot Cr \\ M_f &= M_s - 215 \text{ (error} \pm 15\ ^\circ\text{C)} \end{aligned} \tag{2.2}$$

where C, Mn, Ni, Cr, Si and Mo are the concentrations of the corresponding alloying elements in weight per cent. The above relationships are valid for low alloy steels with less than 0.6% C, 4.9% Mn, 5.0% Cr, 5.0% Ni and 5.4% Mo. Keep in mind that these formulae are empirical based, and the values of M_s and M_f may vary with the change of chemical compositions and concentrations of alloying elements.

2.3.2. Continuous Cooling Transformation Diagram

TTT diagram has only limited application because very few industrial heat treatments are isothermal processes. Most practical heat treatment processes of steels involve the continuous cooling of the alloys rather than maintaining them at a specific temperature to undergo the isothermal transformation indicated by a TTT diagram. Under such conditions, it would be desirable and significant to refer to an alternative kind of transformation diagram to account properly for the microstructures formed from austenite during continuous cooling operations. Continuous cooling transformation (CCT) diagrams of steels determine the nature and amounts of phases in steels on cooling them continuously from a high temperature to room temperature, as practiced in an industrial or normal laboratory situation. Figure 2.7 presents a CCT diagram of a microalloyed steel containing 0.17% carbon. Several cooling curves superimposed on the figure indicate the critical cooling rate, and also how the transformation behaviour and final microstructure are influenced by the cooling rate.

Of interest, the critical cooling rate in a CCT diagram determines the final microstructure that transformed from austenite.

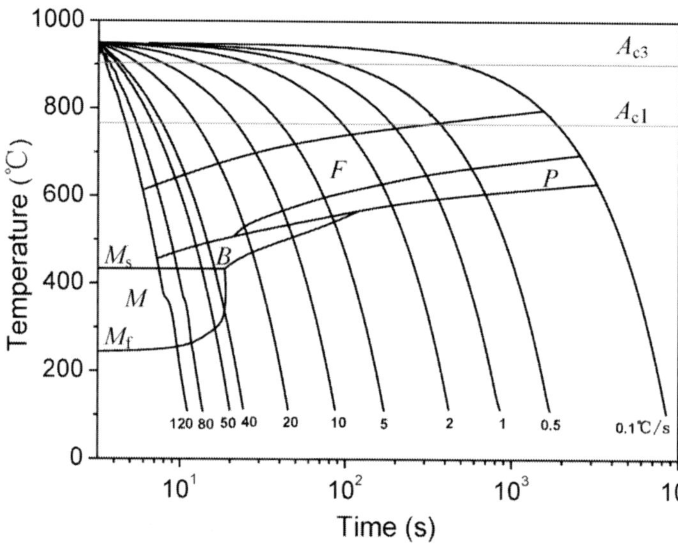

Figure 2.7. CCT diagram of a microalloyed steel containing 0.17% carbon. *F*, *P*, *B* and *M* represent ferrite, pearlite, bainite and martensite respectively.

In Figure 2.7, for example, pearlitic transformation does not occur when the cooling rate is higher than 20°C/s. Bainitic transformation zone appears only above a certain cooling rate of 5°C/s. Martensitic transformation takes place as the cooling rate is higher than 40°C/s. Therefore, to produce a full "ferrite + pearlite" microstructure, it would be necessary to decrease the cooling rate lower than 5°C/s.

The phase transformation behaviour of a microalloyed steel can be displayed as a function of cooling rate on a CCT diagram. CCT diagram can be used to predict the microstructures, and thus the properties, which would result from a specific cooling rate, whether due to fabrication, thermal processing, or welding, and as such are important for engineers and designers in the development, processing and use of microalloyed steels. In forging or rolling, these CCT diagrams can be used to help design the deformation and the subsequent heat treatment processes. Similarly, CCT diagrams serve as a quick reference on cooling rates to avoid the formation of deleterious microstructures, which could be too hard, too soft, susceptible to corrosion and hydrogen-assisted cracking, etc.

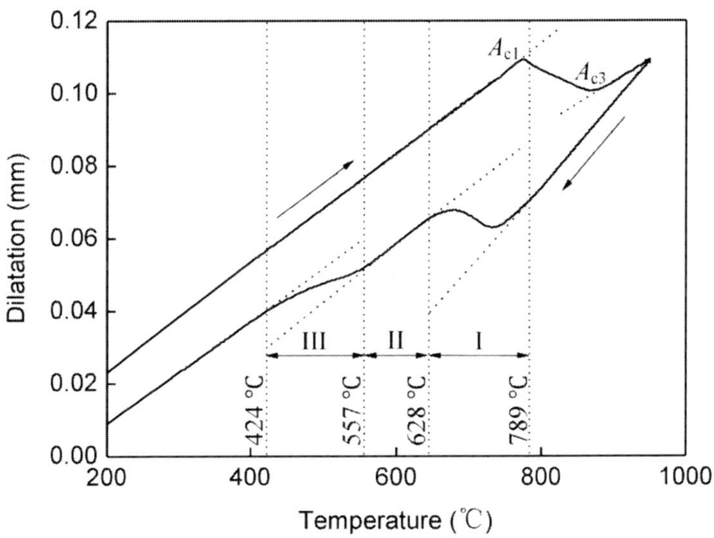

Figure 2.8. Dilatation versus temperature curve of a microalloyed steel with 1% tungsten at the cooling rate of 0.5°C/s. I=Austenite→Ferrite→Pearlite; II=No Transformation; and III=Austenite:Ferrite:Pearlite→Bainite [11].

To determine a CCT diagram, dilatometric method is usually applied. By using this method, the sample (usually cylindrical rod) is continuously cooled from austenite region and the changes in specific volume during phase transformation at different temperatures are simultaneously recorded as a function of time. A typical dilatation versus temperature curve obtained after a dilatometer test of a microalloyed steel with 0.5% tungsten at the cooling rate of 0.5°C/s is shown in Figure 2.8 [11]. The up and down arrows indicate the progress of heating and cooling cycles, respectively. From this curve, the start and finish of a phase transformation can be easily detected. However, as the onset of pearlitic transformation cannot be clearly detected on the dilatation-temperature curve, dilatometric method is generally combined with other techniques such as microscopy, hardness, or others to determine accurately the start and finish of phase transformations.

In this chapter, we will focus on the effect of tungsten on the phase transformation behaviour of microalloyed steels during continuous cooling processes. The isothermal transformation has not been addressed due to its rare application in the practical heat treatment process of these steels.

2.4. Effect of Tungsten on CCT Diagrams

Different phases, such as austenite and ferrite, in steels have distinctly different specific volumes. Hence, it is possible to differentiate the volume change (length change) when phase transformation occurs from the linear thermal expansions during heating and cooling of steels by employing a high resolution dilatometer. To conduct dilatometer tests, cylindrical specimens with a diameter of 3 mm and height of 10 mm were machined from the microalloyed forming steels. Dilatometer specimens were heated to the temperature of 950°C at a rate of 20°C/s, then held for 10 min, and finally cooled to room temperature at different linear cooling rates ranging from 0.1 to 120°C/s. The schedule for dilatometer tests is schematically shown in Figure 2.9 [11]. The CCT diagrams are determined using the method described in Section 2.3.2.

CCT diagrams serve an incredibly useful purpose in representing the transformation characteristics of microalloyed steels which are not isothermally heat treated, and in revealing the role of alloying elements in affecting the microstructures of the steels. In this chapter, three microalloyed steels with different tungsten contents, i.e., 0W, 0.1W and 1W, have been employed to study the effect of tungsten addition on the CCT diagrams. The real chemical compositions are given in Table 1.4 of Chapter 1.

Figure 2.10 presents the CCT diagram of the 0W steel [11]. Such a CCT diagram exhibits consecutive characteristic of the phase transformation and multi-transformation curves, revealing complex microstructures as cooling rate is increased from 0.1 to 120°C/s. Ferrite, pearlite, bainite and martensite are involved in the diagram. Ferrite forms throughout the cooling rate range from 0.1 to 120°C/s. Pearlite is produced under a cooling rate range from 0.1 to

20°C/s. Bainite starts to be produced by increasing the cooling rate to be higher than 5°C/s. Martensite is observed at the cooling rate higher than 40°C/s, and the value of M_f shows a decreasing trend with an increase of cooling rate. M_f is dependent on the cooling rate and can therefore be varied by changing the cooling rate.

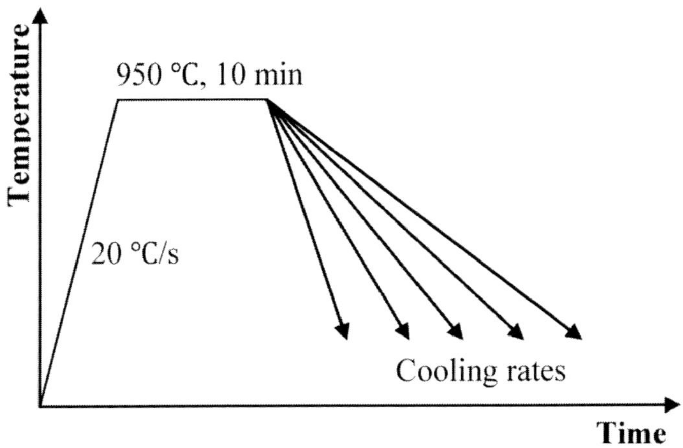

Figure 2.9. Schematic illustration of the schedule for dilatometer tests [11].

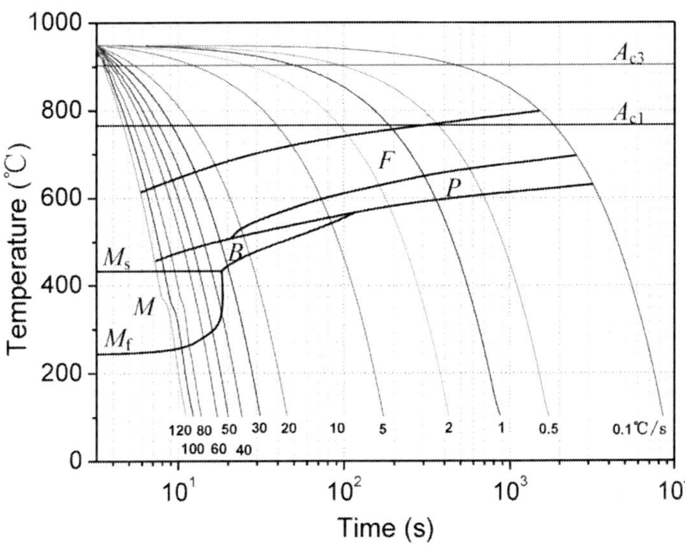

Figure 2.10. CCT diagram of the 0W steel [11].

The shape of the CCT diagram of the 0.1W steel is similar to that of the 0W steel, as shown in Figure 2.11 [11]. The difference is that the ferrite, pearlite, bainite and martensitic transformation ranges in the 0.1W steel are shifted to a longer time side as compared with that in the 0W steel. Moreover, the addition of tungsten narrows the range of pearlitic transformation. Significant effect of tungsten on the CCT diagram is observed when its content is increased to 1%. As shown in Figure 2.12 [11], in the cooling rate range of 0.1 to 1°C/s, the transformation ranges can be classified in the higher temperature range (above 625-640°C) and the lower temperature range (below 538-599°C). The two transformation ranges are separated by a non-transformation range in which no progress of the transformation is observed on cooling. Transformation products are ferrite and pearlite in the higher temperature range, and only bainite in the lower temperature range. Bainitic transformation occurs even though at the very low cooling rate of 0.1°C/s. In addition, as the tungsten content is increased to 1%, a further shift of the ranges of transformation products to the right side of the CCT diagram is caused in contrast to that of the 0.1W steel, as shown in Figures 2.11 and 2.12.

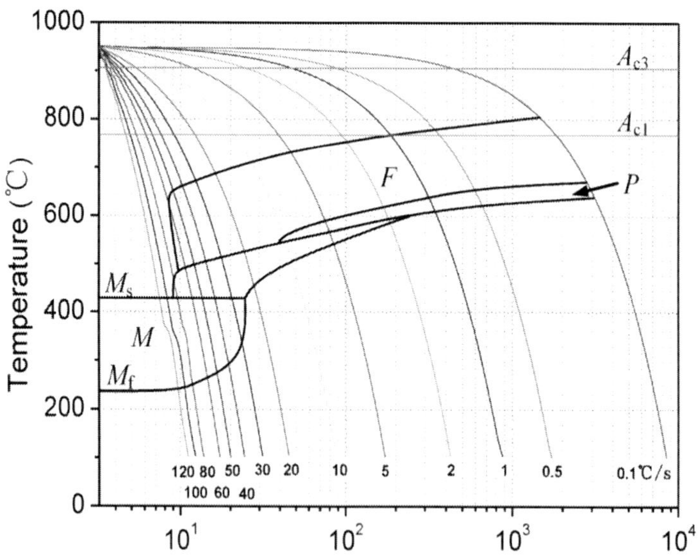

Figure 2.11. CCT diagram of the 0.1W steel [11].

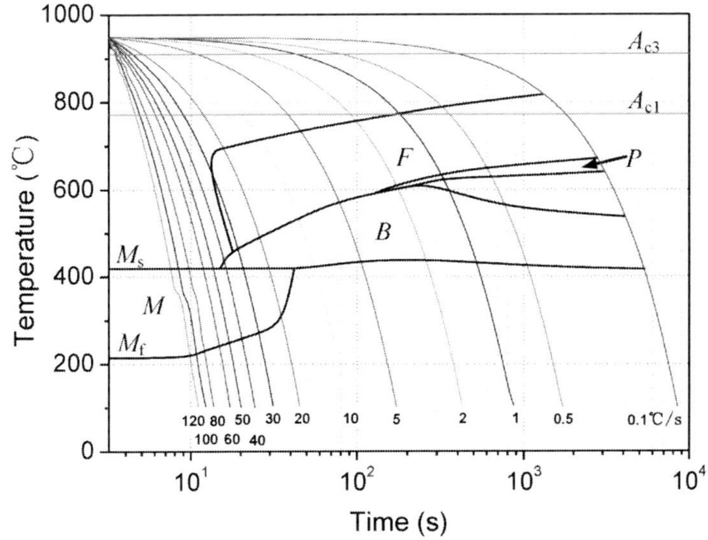

Figure 2.12. CCT diagram of the 1W steel [11].

Table 2.1. Critical cooling rates for phase transformations and obtaining complete ferrite/pearlite [11]

Steels	*F* starting	*P* starting	*B* starting	*M* starting	Complete (*F*/*P*)
0W	-	20°C/s	5°C/s	40°C/s	2°C/s
0.1W	60°C/s	10°C/s	2°C/s	30°C/s	1°C/s
1W	30°C/s	2°C/s	-	20°C/s	-

Based on the CCT diagrams of the 0W, 0.1W and 1W steels, the critical cooling rates for the start of transformations of ferrite, pearlite, bainite and martensite, and for obtaining complete ferrite/pearlite can be obtained, as presented in Table 2.1 [11]. It can be seen that tungsten addition induces decreased critical cooling rates. Ferrite and pearlite are more readily formed in the 0W steel than that in the tungsten-containing steels. Tungsten promotes the formation of bainite and martensite even though at a very low cooling rate. Especially, when the tungsten content is increased to 1%, bainite is formed at the lowest cooling rate of 0.1°C/s, and no complete ferrite/pearlite microstructure is formed in the cooling rate range of 0.1 to 120°C/s.

Tungsten produces both quantitative and qualitative changes in the kinetics of transformation characteristics. The results obtained by Totten [12] revealed that tungsten shifts the transformation ranges to the longer time side of the isothermal transformation diagrams, and two separated transformation ranges in the diagram will be caused as the steel is alloyed with high content of tungsten. This fact is consistent with the results obtained from Figures 2.11 and 2.12 that the transformation ranges are shifted to the right side of the CCT diagram after 0.1% tungsten is added, and two transformation ranges are separated when the tungsten content is increased to 1%.

2.5. EFFECT OF TUNGSTEN ON PHASE TRANSFORMATION TEMPERATURE

Table 2.2 presents the values of A_{c1}, A_{c3} and M_s for the 0W, 0.1W and 1W steels. It is clear that tungsten addition induces increased A_{c1} and A_{c3} temperatures. Differently, tungsten shows a positive effect on decreasing the starting temperature of martensitic transformation. Figure 2.13 shows the variation of the starting temperatures of austenite-to-ferrite (A_{r3}) as a function of cooling rates [11]. It can be seen that the addition of tungsten increases A_{r3} and shifts the A_{r3} temperature to lower cooling rates.

Table 2.2. Values of A_{c1}, A_{c3} and M_s for the 0W, 0.1W and 1W steels

Steels	A_{c1}	A_{c3}	M_s
0W	766°C	903°C	434°C
0.1W	767°C	906°C	429°C
1W	771°C	910°C	420°C

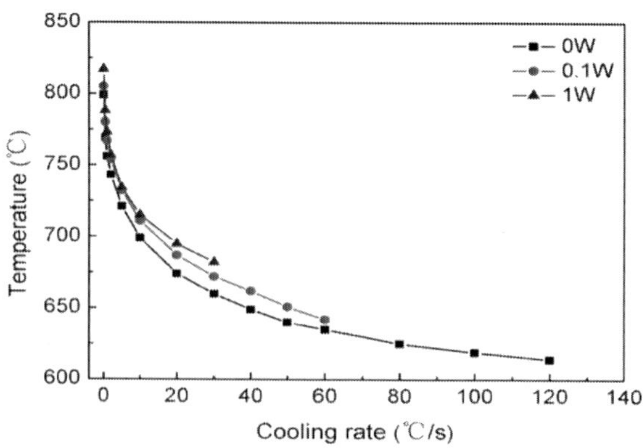

Figure 2.13. Dependence of A_{r3} on cooling rate [11].

As a strong ferrite former, tungsten effectively stabilise ferrite by raising the ferrite to austenite transformation temperature. The ferrite field will be expanded as tungsten is added. As a result, the A_{c1} and A_{c3} temperatures during heating, and the A_{r3} temperature during cooling show increasing trends with tungsten. M_s and the factors that affect its value have been broadly investigated. M_s is a definite temperature for a specified steel. M_s strongly depends on the chemical compositions of the steels. Several studies have indicated that tungsten depresses the M_s temperature [13,14]. Austenite grain size also plays an important role on affecting M_s. It is well accepted that refined austenite grain size induces decreased M_s [15,16]. Increase of tungsten content results in finer austenite grain size [11,17], which in turn decreases the M_s temperature. It has been suggested that the effects of different alloying elements, such as tungsten, on M_f are similar in magnitude to their effect on M_s [18,19], i.e., M_f will decrease when M_s is decreased.

2.6. Effect of Tungsten on Transformed Microstructure

The transformed microstructures of the 0W, 0.1W and 1W steels after continuous cooling are presented in Figure 2.14 [11]. The label at the upright corner of each image indicates the cooling rate. The microstructure of the 0W steel at the cooling rate of 0.5°C/s is composed of ferrite and pearlite (Figure 2.14a). When the cooling rate is increased to 30°C/s, the microstructure becomes a mixture of ferrite and bainite (Figure 2.14b).

At the highest cooling rate of 120°C/s, the microstructure comprises of martensite, bainite and small quantity of ferrite, and predominately martensite, as shown in Figure 2.14c. After 0.1% tungsten addition, the microstructure of the 0.1W steel is still composed of ferrite and pearlite when the cooling rate is 0.5°C/s (Figure 2.14d). As the cooling rate increases, the 0.1W steel exhibits a mixed microstructure of ferrite, bainite and martensite at the cooling rate of 30°C/s (Figure 2.14e), and then a fully martensitic microstructure at the cooling rate of 120°C/s (Figure 2.14f). The addition of 1% tungsten shows significant effect on the microstructures after continuous cooling in contrast to the 0W and 0.1W steels. At the cooling rate of 0.5°C/s, bainite is observed in the microstructure of the 1W steel in addition to ferrite and pearlite (Figure 2.14g). As the cooling rate is increased to 30°C/s, the microstructure of 1W steel is mainly composed of martensite, and small quantity of bainite and ferrite (Figure 2.14h). At the highest cooling rate of 120°C/s, the 1W steel exhibits a full martensite microstructure, as shown in Figure 2.14i. The microstructural observation obtained from Figure 2.14 reveals that tungsten plays an important role in obtaining different microstructural combinations during continuous cooling process.

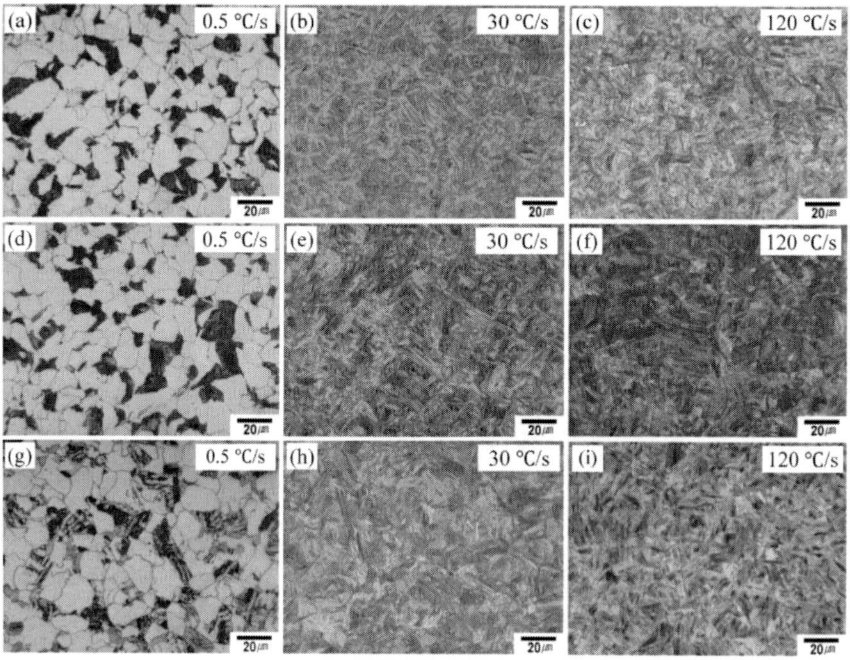

Figure 2.14. Microstructures of the continuously-cooled steels: (a,b,c) 0W, (d,e,f) 0.1W, and (g,h,i) 1W [11].

2.7. Effect of Tungsten on Hardness after Phase Transformation

The relationship between hardness and cooling rate is presented in Figure 2.15 [11]. The addition of tungsten brings about an increase in the hardness at a given cooling rate. For the 0W and 0.1W steels, the hardness shows a gradually increasing trend with an increase of cooling rate. For the 1W steel, however, the hardness increases significantly in the cooling rate range of 0.1 to 40°C/s, then increases slightly when the cooling rate is higher than 40°C/s.

In order to describe the change of hardness with cooling rate, an exponential equation is developed:

$$HV = \beta_1 + \beta_2 \cdot \exp(\beta_3 \cdot v_c) + \beta_4 \cdot \exp(\beta_5 \cdot v_c) \quad (2.3)$$

where HV is Vickers hardness (HV2), v_c is the cooling rate (°C/s), and β_1, β_2, β_3, β_4 and β_5 are constants. The quantitative expressions between HV and v for the 0W, 0.1W and 1W steels can be acquired through regression analysis, and the adjust coefficient of determination R^2_{adj} can be obtained to evaluate the performance of the calculated equations. The calculated values of β_1, β_2, β_3, β_4 and β_5, as well as relevant R^2_{adj}, are given in Table 2.3 for each steel condition [11].

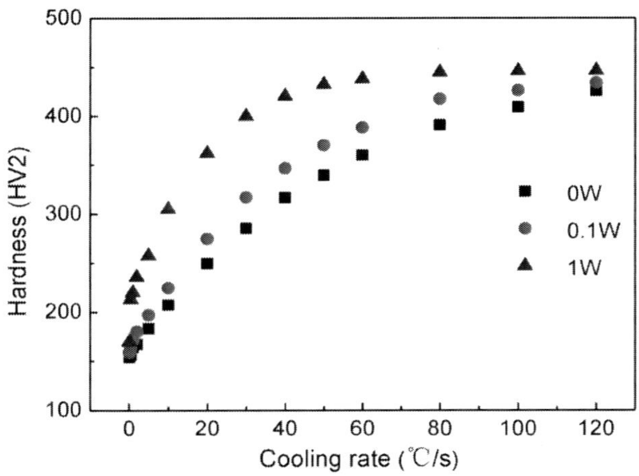

Figure 2.15. Dependence of hardness on cooling rate of the 0W, 0.1W and 1W steels [11].

Table 2.3. Calculated parameters for the exponential equation [11]

Steels	β_1	β_2	β_3	β_4	β_5	R^2_{adj}
0W	454.949	-50.603	-0.019	-248.383	-0.019	0.994
0.1W	453.420	-47.542	-0.025	-244.496	-0.025	0.992
1W	449.360	-241.528	-0.051	-1073.944	-33.101	0.999

Figure 2.16 presents the histograms of residual diagnosis results for the 0W, 0.1W and 1W steels after regression analysis [11]. It can be seen that all the residuals are normally distributed; indicating accurate calculation results can be obtained. Figure 2.17 shows the comparative plots for the tested and calculated hardness values at different cooling rates [11]. It can be seen that the results of calculation agree well with the experimental results.

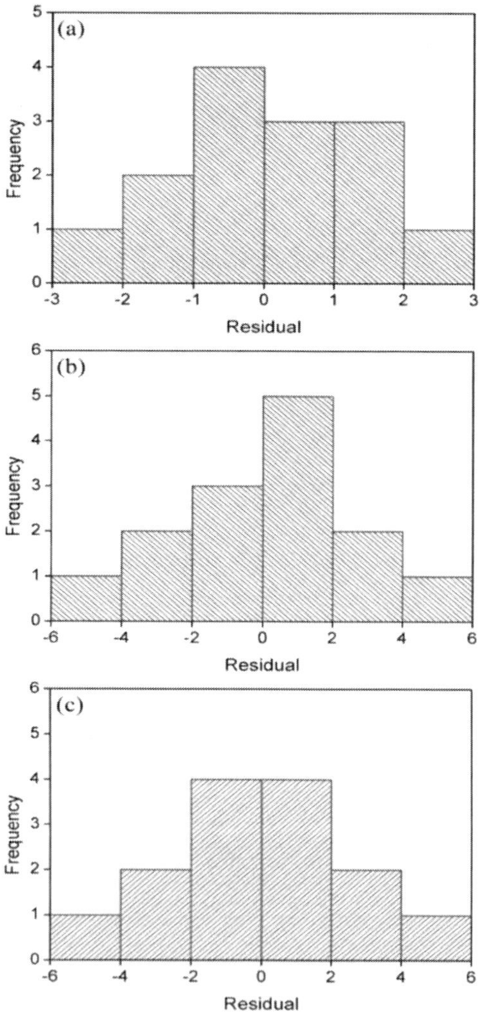

Figure 2.16. Histogram of residual diagnosis results for the (a) 0W, (b) 0.1W and (c) 1W steels after regression analysis [11].

The hardness of steels is strongly related to the chemical compositions and final microstructures. Tungsten addition in microalloyed steels has a beneficial effect on the improvement of hardness [20]. With an increase of the applied cooling rate, the transformed structure evolves from ferrite+pearlite/bainite (0W and 0.1W: ferrite+pearlite; 1W: ferrite+pearlite+bainite) to final martensite.

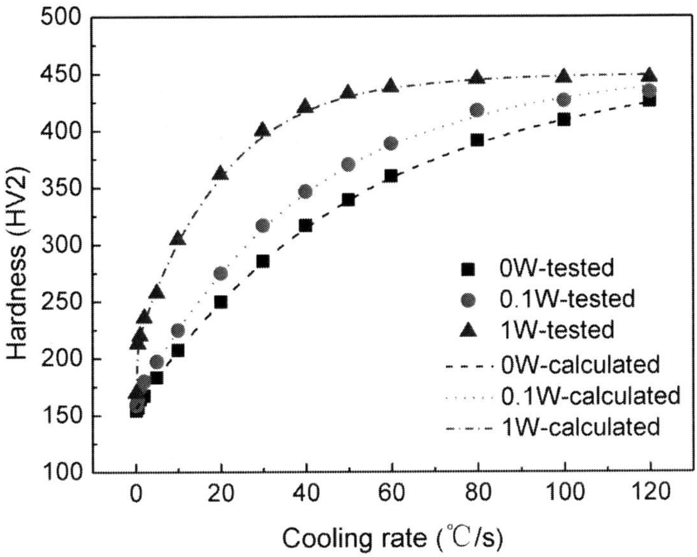

Figure 2.17. Comparison between the tested and calculated hardness values at different cooling rates [11].

The content of bainite and martensite in the final microstructure gradually increases as the cooling rate increases, as shown in Figures 2.10, 2.11, 2.12 and 2.14. Consequently, the hardness increases with an increase of the cooling rate due to the changes in the microstructural combination [21].

2.8. Summary

This chapter describes the iron-carbon equilibrium diagram and the phase transformations that commonly occur in microalloyed steels. The isothermal and CCT diagrams are introduced with purpose of providing an in-depth understanding of the phase transformation mechanism. CCT characteristics of microalloyed steels with different tungsten contents (0, 0.1 and 1 wt.%) are systematically investigated with the aim of obtaining necessary information for application to practical processes, such as heat treatment of these steels. CCT diagrams of microalloyed steels with different tungsten contents are created and presented by varying the cooling rates from 0.1 to 120°C/s. The effects of tungsten on critical cooling rate for phase transformation and microstructural components, phase transformation temperature, transformed microstructure, and hardness after phase transformation are analysed in detail.

The ranges of transformation products are shifted to the right side of the diagram when the tungsten content increases. The shape of the CCT diagram of the 0.1W steel is similar to that without tungsten. The addition of 1% tungsten induces two separated transformation ranges in the cooling rate range of 0.1 to 1°C/s in the diagram. Both the austenitisation starting and finishing temperatures during heating are raised after tungsten addition. The critical cooling rates for phase transformations and obtaining complete ferrite+pearlite microstructures decrease as the tungsten content is increased. The martensitic transformation temperature is decreased after tungsten addition. The addition of tungsten leads to an increased level of hardness, and the hardness obeys an exponential type relationship with cooling rate.

REFERENCES

[1] H.K.D.H. Bhadeshia, R.W.K. *Honeycombe, Steels: Microstructures and Properties*, 3rd ed., Elsevier, 2006.

[2] R.C. Sharma, *Principles of Heat Treatment of Steels*, New Age International, 1996.

[3] H.I. Aaronson, C. Laird, K.K. Kinsman, *Phase Transformations*, American Society for Metals, Metals Park, Ohio, 1970.

[4] H.K.D.H. Bhadeshia, *Bainite in Steels: Transformations, Microstructure and Properties*, 2nd ed., The University Press, Cambridge, 2001.

[5] Z. Nishiyama, *Martensite Transformation*, Elsevier, 2012.

[6] K.W. Andrews, Empirical formulae for the calculation of some transformation temperatures, *Journal of the Iron and Steel Institute*, 203 (1965) 721-727.

[7] R.A. Grange, H.M. Stewart, The Temperature Range of Martensite Formation, *Transactions of the AIME*, 167 (1946) 467-490.

[8] G. Krauss, *Principles of Heat Treatment and Processing of Steels*, ASM International, 1990.

[9] A.V. Sverdlin, A.R. Ness, The effects of alloying elements on the heat treatment of steel, in: G.E. Totten, M.A.H. Howes (Eds.), *Steel Heat Treatment Handbook*, Marcel Dekker, New York, 1997.
[10] W. Steven, A.G. Haynes, The temperature of formation of martensite and bainite in low alloy steels, *Journal of the Iron and Steel Institute*, 183 (1956) 349-359.
[11] J. Zhao, Z. Jiang, J.S. Kim, C.S. Lee, Effects of tungsten on continuous cooling transformation characteristics of microalloyed steels, *Materials and Design*, 49 (2013) 252-258.
[12] G.E. Totten, *Steel Heat Treatment: Metallurgy and Technologies*, 2nd ed., CRC Press, 1997.
[13] P. Payson, C.H. Savage, Martensite reactions in alloy steels, *Transactions ASM*, 33 (1944) 261-275.
[14] E.S. Rowland, S.R. Lyle, The application of M_s points to case depth measurement, *Transactions ASM*, 37 (1946) 27-46.
[15] A. García-Junceda, C. Capdevila, F.G. Caballero, C. García de Andrés, Dependence of martensite start temperature on fine austenite grain size, *Scripta Materialia*, 58 (2008) 134-137.
[16] H.S. Yang, H.K.D.H. Bhadeshia, Austenite grain size and the martensite-start temperature, *Scripta Materialia*, 60 (2009) 493-495.
[17] J. Zhao, T. Lee, J.H. Lee, Z. Jiang, C.S. Lee, Effects of tungsten addition on the microstructure and mechanical properties of microalloyed forging steels, *Metallurgical and Materials Transactions A*, 44A (2013) 3511-3523.
[18] W. Steven, A.G. Haynes, The temperature of formation of martensite and bainite in low-alloy steels - some effects of chemical composition, *Journal of the Iron and Steel Institute*, 183 (1956) 349-359.
[19] G.F. Vander Voort, *Atlas of Time-Temperature Diagrams for Irons and Steels*, ASM International, Ohio, 1991.
[20] J. Zhao, Z. Jiang, C.S. Lee, Effects of tungsten addition and heat treatment conditions on microstructure and mechanical properties of microalloyed forging steels, *Materials Science and Engineering A*, 562 (2013) 144-151.
[21] A. Calik, Effect of cooling rate on hardness and microstructure of AISI 1020, AISI 1040 and AISI 1060 steels, *International Journal of Physical Sciences*, 4(9) (2009) 514-518.

Chapter 3

EFFECT OF TUNGSTEN ON PRECIPITATION

ABSTRACT

Precipitates in microalloyed steels play important roles in the grain size refining and precipitation hardening. In this chapter, the roles of precipitates in both austenite and ferrite phases of microalloyed steels are first described. Typical investigations on the precipitates in various microalloyed steels are then briefly reviewed. The dissolution and precipitation reactions of precipitates are discussed in order to understand the precipitation behaviour of various precipitates. The solubility products of precipitates in microalloyed steels are summarised. The classical nucleation theory is introduced with the aim of understanding the nucleation process of precipitated phases in microalloyed steels. The growth of precipitated phases is described in the form of growth rate. In order to study the stability of precipitates such as carbides, nitrides and/or carbonitrides in microalloyed steels at elevated temperatures, the existing status of precipitates at different reheating temperatures has been identified. The morphology and structure of precipitates at different reheating temperatures are analysed. The effects of tungsten addition on the shape, size and amount of precipitates in the microalloyed forging steels are discussed. The concentration of tungsten in precipitates with different shapes is quantified.

3.1. PRECIPITATES IN MICROALLOYED STEELS

An important character of microalloyed steels as compared to other steel grades is the use of a small amount of microalloying elements to obtain a good combination of excellent toughness and strength through grain size controlling and precipitation strengthening. The added microalloying elements such as vanadium, niobium and titanium in microalloyed steels are generally precipitates (e.g., carbides, nitrides or carbonitrides) forming elements. Optimised microstructure and enhanced mechanical properties will be available through the control of precipitation behaviour of various kinds of fine-sized particles during heat treatment and hot working.

Precipitates play roles in both austenite and ferrite phases of microalloyed steels. Precipitates in austenite phase have two main effects. Firstly, finely dispersed precipitates impose pinning forces that restrict the austenite grain boundary movement and grain growth, allowing a successively finer austenite grain size to be obtained. Austenite is the parent phase of all other final microstructures in microalloyed steels, and the microstructures of austenite transformation products depend on austenite grain size. Refinement of austenite grain size in microalloyed steels assists in the production of the final refined ferrite grain size with enhanced mechanical properties. Secondly, precipitates at dislocations in austenite may retard the occurrence of recrystallisation. Consequently, a pancaked austenite grain structure will be formed, which results in promoting the refinement of ferrite grain size during the subsequent austenite-to-ferrite transformation. These two effects are usually used in combination to reduce the austenite grain size before austenite-to-ferrite transformation during heat treatment and/or hot working processes of microalloyed steels.

In ferrite, precipitates generally take effect in the form of precipitation strengthening. Particles may precipitate in ferrite grain or at the migrating austenite/ferrite interface during the austenite-to-ferrite transformation. Improved yield strength of microalloyed steels can be acquired by increasing the resistance to dislocation motion in the presence of finely dispersed precipitates.

Microalloyed steels contain many kinds of precipitates forming elements such as aluminium, vanadium, niobium, titanium, molybdenum and chromium etc. Research on the precipitation behaviour in microalloyed steels is a challenging task due to the different chemical and physical features of each microalloying element, and their combination effects in the forming of complex precipitates. Since each precipitate former has specific precipitation preferences, they should be considered separately, including their effects on processing and the resulting structure-property relationships. So far however, precipitation in microalloyed steels is still not fully understood.

3.2. Investigations on Precipitates in Microalloyed Steels

Extensive investigations on the precipitates in various microalloyed steels have been conducted by many researchers. In this section, typical research findings in the respects of identification of precipitates, dissolution of precipitates, evolution of precipitates in reheating, cooling, casting/solidification and thermomechanical controlled processing (TMCP), and thermodynamic and kinetic modelling of precipitation behaviour in microalloyed steels are briefly reviewed.

3.2.1. Identification of Precipitates

Identification of precipitates in microalloyed steels is of importance for the process control with the purposes of modifying the microstructure and improving the mechanical properties due to the beneficial roles that precipitates played in the steels.

Davis and Strangwood [1] prepared carbon extraction replicas from three niobium microalloyed steels, and determined fine precipitate size distribution and areal number density in ferrite and pearlite. One problem is that carbon extraction replicas have been found not to be effective in revealing very fine (less than 5 nm) particles [2]. Sha et al. [3] employed a carbon extraction replica method to identify the precipitates in a niobium-vanadium-titanium microalloyed steel, and found that niobium and titanium carbonitrides formed at austenite grain boundaries and inside austenite grains could be clearly observed by using transmission electron microscope (TEM). However, the number of precipitates being revealed was greatly reduced by using this developed method as compared to the conventional approach due to the dissolution of a certain amount of particles in etching with longer time and higher temperature involved. Pischow et al. [4] compared scanning tunnelling microscopy (STM) with TEM in the study of precipitates in a microalloyed steel. In the STM method, samples were mechanically polished and then etched with a 1% nital solution. TEM samples were prepared by electropolishing at subzero temperature in a solution of 75% CH_3OH and 25% HNO_3 with voltage of 8 V and time of 1 to 2 min. They claimed that the sizes of the precipitates measured by STM and TEM techniques were nearly equal, and similar precipitate clusters could be observed by both STM and TEM. The preparation of STM samples, however, must be done carefully even though the preparation methods for STM are generally easier than that for TEM because the STM images are based on the topography, the surface, its quality and the preparation used. Wiskel et al. [5] successfully applied a small angle neutron scattering (SANS) method to quantify the precipitate characteristics including average precipitate size, number of precipitates and distribution of fine precipitates in X-70 and X-80 pipeline steels and in grades 80 and 100 microalloyed steels. SANS is versatile in the quantification of the size distribution of precipitates with the size on the order of 1 to 100 nm. The challenges associated with applying SANS to commercially produced microalloyed steels include (i) the low volume fraction of the very fine precipitates, and (ii) the wide range of precipitate sizes resulting from various precipitation events.

Rentería-Borja et al. [6] found that use of atomic force microscopy (AFM) was effective to quantify the nanometric precipitates in both niobium and vanadium-molybdenum microalloyed steels. In contrast with TEM, quantification with AFM could be achieved in large fields without the hassle of complicated sample preparation procedures. The results obtained using AFM were found to be consistent with similar TEM reports. He and Baker [7] proposed an aluminium replica extraction method to study the effects of 0.016 wt.% zirconium on the precipitates in a titanium-niobium microalloyed HSLA steel. Precipitates were analysed using a TEM fitted with an energy-dispersive analysis of X-rays (EDAX), and the lattice parameters of the precipitates were obtained from the selected area election diffraction patterns (SAED). The results indicated that complex zirconium-containing oxides and titanium-zirconium and titanium-niobium-zirconium nitrides and/or carbonitrides could be well identified by using the proposed method. For quantifying fine precipitates, Lu et al. [8] developed two matrix dissolution techniques, i.e., electrolytic and chemical dissolutions, to extract precipitates from a grade 100 microalloyed steel. By using the techniques developed, relatively large volumes of steel could be analysed, and statistically significant quantities of precipitates with different sizes could be collected.

3.2.2. Dissolution of Precipitates

Park and Lee [9] proposed an approach for the determination of Nb(C,N) dissolution temperature by electrical resistivity measurement in a low carbon (0.021 wt.%) microalloyed steel after solution treatment at temperatures ranging from 1150 to 1250°C. Electrical resistivity is one of the physical characteristics of a steel and is dependent on temperature and crystal defects such as solute atoms, dislocations and impurities. They found that the dissolution temperature of Nb(C,N) was about 1200°C for a holding time of 10 min. Hong et al. [10] studied the dissolution behaviour of precipitates in niobium and niobium-titanium microalloyed steels by means of carbon extraction replicas prepared from the deformed and water-quenched specimens after austenitising treatment at 1250°C for 400 s. They claimed that niobium was not fully soluble at the reheating temperature of 1250°C in the niobium-titanium microalloyed steel, and could be still remained in titanium-rich (Ti,Nb)(C,N) carbonitrides. Differently, NbC carbides or Nb(C,N) carbonitrides had completely dissolved at the reheating temperature of 1250°C in the niobium-microalloyed steel due to the fact that the equilibrium solution temperature of NbC or Nb(C,N) is about 1194°C according to the equation given by Irvine et al. [11].

Medina [12] described a method for the determination of precipitation-time-temperature (PTT) diagrams for three niobium, vanadium and titanium microalloyed steels by holding torsion-deformed specimens at 1230°C for 10 min and then cooling them rapidly to the testing temperature, and claimed that all the niobium and vanadium precipitates were completely dissolved while titanium nitrides were only partially dissolved at this temperature. Yuan and Liang [13] investigated the dissolving behaviour of precipitates in a low carbon niobium-titanium microalloyed steel by isothermally holding at 1300°C for different time before quenching into water, and found that large-sized precipitates containing niobium and titanium still remained even if the steel was held at such high temperature for 48 h. They concluded that the precipitates formed during solidification were (Nb,Ti)(C,N) rather than TiN. The work of Hong et al. [14] on the dissolution behaviour of precipitates in a niobium-titanium-vanadium microalloyed steel by means of aluminium extraction replicas indicated that dendritic niobium-rich (Ti,Nb)(C,N) carbonitrides in the as-cast slab steel had a thermodynamically stable chemistry at temperature lower than 1000°C. As the reheating temperature was increased from 1050 to 1400°C, niobium content within the carbonitrides decreased, while titanium content did not change to 1250°C. Above 1300°C, titanium started to dissolve from the carbonitrides into the austenite matrix. Vanadium carbides/nitrides have the lowest dissolution temperatures as compared to niobium and titanium precipitates.

The dissolution temperatures of some general precipitates determined by Narita [15] and Turkdogan [16] were as follows: VC 777°C, VN 974°C, NbC 1095°C, NbN 1114°C, TiC 1068°C, and TiN 1550°C.

3.2.3. Evolution of Precipitates

In this section, the effects of reheating temperature, cooling rate, casting/solidification rate, and TMCP on the evolution of precipitates in microalloyed steels are described.

- **Effect of reheating temperature.** As described in Section 3.2.2, there are different dissolution temperatures among different kinds of precipitates. Therefore, reheating temperature would have significant effect on the state and characteristic of precipitates in microalloyed steels.

The work of Vervynckt et al. [17] on a niobium microalloyed steel showed that a small amount of 0.01 wt.% precipitated niobium was found, and most precipitates were identified as (Nb,Ti)(C,N) and (Nb,Ti)C with an average size of 60 nm after reheating at 1250°C for 5 min. Hong et al. [14] found that the dendritic carbonitrides resulted from non-equilibrium cooling were dissolved and then re-precipitated as cubic shape particles along austenite grain boundaries in the reheating temperature range of 1100-1400°C and within austenite grains in the reheating temperature range of 1100-1250°C. They concluded that stable titanium- and nitrogen-rich (Ti,Nb)(C,N) precipitated primarily along austenite grain boundaries, followed by precipitation of titanium- and carbon-rich (Ti,Nb)(C,N) within austenite grain during reheating process. Zhuo et al. [18] investigated the evolution of precipitates during reheating process and found that titanium-rich (Ti,Nb)(C,N) was stable at about 1200°C and niobium-rich (Ti,Nb)(C,N) was stable at about 800°C. Particles of (Ca,Mn)S and TiN precipitated from Ca-Al oxide after reheating at 1200°C for 1 h, while Ca-Al-Ti oxide existed stably during the reheating process at 800°C for 1 h. Gündüz and Cochrane [19] claimed that the holding time during reheating process affected the formation of specific precipitate AlN, i.e., reheating at 900°C for 26 h promoted the formation of AlN while it could not be observed after reheating at this temperature for only 1 h.

H.R. Wang and W. Wang [20] found that reheating temperature affected significantly the size and morphology of precipitates. Their results showed that the size of cuboidal carbonitrides was reduced from 40 to 20 nm as the reheating temperature was decreased from 1200 to 1150°C. The morphology of carbonitrides in the HAZ was transformed from cuboidal rectangular shape with the largest edge length of exceeding 500 nm when a niobium-titanium bearing steel was reheated at 1150°C, but the morphology was remained basically unchanged when the steel was reheated at 1150°C even though the size of complex carbonitrides had increased in the HAZ as compared to that in the base steel.

- **Effect of cooling rate.** The precipitation behaviour of precipitates is affected by the cooling rate when cooling a microalloyed steel from a reheating temperature. Ma et al. [21] studied the effect of different cooling rates on the precipitation behaviour of carbonitrides in a microalloyed steel slab by confocal laser scanning microscopy. They found that the precipitates in the steel were coarse and mostly irregularly shaped when the cooling rate was less than 3 K/s, and both the volume fraction and the size of precipitates exhibited a reducing trend with the increase of cooling rate. The work of Sha et al. [22] on a niobium-vanadium-titanium microalloyed steel showed that interphase precipitation appeared at the cooling rates of 1 to 5°C/s. The volume fraction of general precipitates under the size of 10 nm reduced with the increase of cooling rate at the simulated coiling temperatures of 650 and 600°C. Lower coiling temperature would induce more precipitates being generated after hot rolling under the same cooling rate. Zhuo et al. [18] found that titanium-rich (Ti,Nb)(C,N) precipitates were transformed gradually to niobium-rich (Ti,Nb)(C,N) precipitates with an increase of the cooling rate. The results obtained by Gündüz and Cochrane [19] indicated that a fast cooling rate refined the size of precipitates, and the precipitates would be coarsened as the cooling rate was decreased.
- **Effect of casting/solidification rate.** The size of precipitates in microalloyed steels is related with the casting/solidification rate in a casting process. Longauerová et al. [23] studied the effect of casting rate on the precipitation behaviour in a titanium-niobium microalloyed steel. At a lower casing rate of 0.5 m/min, the size of precipitates was found to distribute non-uniformly, and coarse globular precipitates with size up to 76.9 nm and angular precipitates

with size up to 107.7 nm precipitated during the casing process. At a higher casting rate of 0.8 m/min, however, the precipitates appeared to be similar in shape, and the size was significantly refined as compared to that at a lower casting rate. Xu et al. [24] studied the precipitation behaviour in continuous casting of a microalloyed steel under different casting rates, and found that an increase in casting rate greatly reduced the amount of precipitates, and was therefore beneficial in preventing surface cracks caused by the thermal stresses in the spray-cooling zones and mechanical stresses from unbending if precipitates were present. The work of Arribas et al. [25] indicated that a faster solidification rate, i.e., a faster cooling rate, in casting brought about a finer precipitation of TiN particles which were very effective in controlling austenite grain growth in the subsequent hot working processing.

- **Effect of TMCP.** TMCP of microalloyed steels is a metallurgical process that integrates deformation and heat treatment into a single hot rolling/forging process. The purpose of TMCP of microalloyed steels is to exploit the effect of plastic deformation above and below the non-recrystallisation temperatures on the microstructure of austenite with the purpose of developing a most favourable, fine-grained microstructure on transformation to achieve improved mechanical properties by precipitation, solid-solution strengthening, dislocation strengthening and refinement of ferrite grain size. The precipitation process during TMCP is complex, because it occurs within the steel matrix which evolves continuously from the austenite to ferrite. Many factors including steel compositions, processing histories, strain applied etc. affect the precipitation behaviour in microalloyed steels [26].

Misra et al. [27] compared the precipitation behaviour of copper and fine-scale carbides during TMCP and isothermal aging of copper-bearing niobium microalloyed high strength steels, and found that ε-copper precipitation took place in polygonal ferrite and at the austenite/ferrite interface during TMCP, whereas during isothermal aging precipitation of ε-copper took place at dislocations. Fine-scale carbides and ε-copper precipitates were mutually exclusive and they were not synergistic. The precipitation of carbides did not influence copper precipitation and vice versa. The work of Wu et al. [28] indicated that precipitation occurred at austenite grain boundaries and deformation bands at the initial stages of nucleation, and the quantity and size of V(C,N) precipitates increased with the extension of inter-pass time during TMCP. Li et al. [29] studied the precipitation behaviour of V(C,N) precipitates in vanadium microalloyed steels during TMCP. The results showed that the precipitation of V(C,N) particles in the undeformed austenite was very sluggish, and no V(C,N) was precipitated when holding time was less than 780 s. Differently, the precipitation of V(C,N) was significantly promoted after deformation. Pereloma et al. [30] conducted a research on the effect of TMCP conditions on niobium clustering and precipitation in a niobium-titanium microalloyed steel. They claimed that a decrease in the deformation temperature increased the niobium-rich clustering and precipitation in austenite as a result of a slow dislocation annihilation rate and an increased number of cluster/precipitate nucleation sites. Conversely, the extent of clustering and precipitation decreased in ferrite, following depletion of Nb from the matrix. Vervynckt et al. [17] found that the precipitation states of microalloyed steel were greatly affected by several factors including reheating, deformation and cooling during TMCP, and precipitation affected the occurrence of austenite recrystallisation, i.e., recrystallisation occurred during precipitate nucleation and coarsening stages, while it was halted completely during precipitation growth stage.

3.2.4. Modelling of Precipitation

- **Thermodynamic modelling.** Thermodynamic analysis of precipitation is an effective approach for optimising both chemistry and process design of microalloyed steels due to the important roles of precipitates in modifying the microstructure and improving the overall properties of microalloyed steels.

Xu et al. [24] established a thermodynamic model for predicting the occurrence and stability of carbides, nitrides, sulphides and oxides in microalloyed steels, as well as their equilibrium compositions, based on satisfying solubility limits including Wagner [31] interaction between elements, mutual solubility between precipitates, and mass conservation of alloying elements. This model predicted the compositions and amounts of stable precipitates for multicomponent microalloyed steels in liquid, ferrite and austenite phases at any temperature.

In order to analyse the solubility of precipitates in microalloyed steels, Gorbachev and Popov [32] constructed a self-consistent thermodynamic model of the Fe-Nb-C-N system and its constituent ternary subsystems using CALPHAD method. With the aid of the thermodynamic model constructed, phase equilibrium in these systems could be calculated, and the solubility of carbides, nitrides and carbonitrides in the Fe-Nb-C, Fe-Nb-N and Fe-Nb-C-N systems could be well analysed. For the Fe-Nb-C-N system, Luo et al. [33] also proposed a thermodynamic model based on the regular solution sublattice model, which was developed to predict the austenite/carbonitride equilibrium. Experimental verification exhibited acceptable reliability of this developed model. Xiang et al. [34] conducted a thermodynamic research on precipitates in a niobium-titanium microalloyed steel (Fe-Nb-Ti-C-N system). This model could be used to represent the precipitation behaviour of niobium and titanium carbonitrides in low carbon microalloyed steels produced by compact strip production. A thermodynamic model was proposed by Speer et al. [35] to predict the austenite/carbonitride equilibrium in the Fe-Nb-V-C-N system using published solubility data and the Hillert/Staffansson model for stoichiometric phases. This model could be used to estimate equilibrium austenite and carbonitride compositions, and the amounts of each phase, as a function of steel composition and temperature. The model also provided a method to estimate the carbonitride solution temperatures for different steel compositions.

Liu and Jonas [36] developed a model to predict the equilibrium mole fractions of precipitates Ti(C,N), MnS and $Ti_4C_2S_2$ in four microalloyed steels with different titanium additions. The solubility products are calculated from standard Gibbs energies and account for the interaction between alloying elements and the mutual solubility of Ti(C,N). The precipitation of complex vanadium carbonitrides and aluminium nitrides in C-Al-V-N microalloyed steels was discussed by Gao and Baker [37]. They used two thermodynamic models proposed by Adrian [38] and Rios [39], and produced similar results.

Zeng and Wang [40] integrated thermodynamics and kinetics into a single thermo/kinetics computational model to predict the precipitation behaviour of complex precipitates in Nb-Ti microalloyed steels under hot deformation condition. In this model, the equilibrium concentration of substitutional elements in austenite and the driving force for precipitation were calculated thermodynamically, and the time dependence of volume fraction and mean radius of precipitates was predicted kinetically on the basis of classical nucleation and growth theory. Validation by the experimental data in the literature showed a good predictability of the developed model.

- **Kinetic modelling.** The formation, growth, volume fraction and size distribution of second phase precipitates, e.g., carbides, nitrides and/or complex carbonitrides, affect greatly the microstructure and mechanical properties of microalloyed steels. An in-depth understanding of the precipitation kinetics is of significance in controlling the precipitation behaviour in the production process of microalloyed steels with the purpose of obtaining modified microstructural constituents and improved properties.

In order to quantitatively and accurately measure the precipitation kinetics in microalloyed steels, complete dissolution of the pre-existing precipitates formed during any previous fabrication process, such as continuous casting, is necessary. By using electrical resistivity, interrupted compression and electrolytic dissolution methods as well as TEM, Park et al. [41] conducted an experimental research on the dissolution and precipitation kinetics of Nb(C,N) particles in both strain-free and strained austenite region of a low carbon niobium-microalloyed steel, and constructed relevant PTT diagrams of Nb(C,N) precipitation. The results indicated that the interrupted compression and electrical resistivity methods showed more reasonable strain-induced precipitation kinetics as compared to the electrolytic dissolution method, and the electrical resistivity method had wider applications to the dissolution and precipitation kinetics in microalloyed steels. Wang et al. [42] investigated the kinetics of strain induced precipitation of TiC in three titanium-microalloyed steels containing different levels of manganese by using two-stage interrupted compression tests, and successfully obtained PTT diagrams through analysing the softening kinetics curves of deformed austenite. They found that the PTT diagrams constructed showed good agreement with those obtained by the previous researchers. Maugis and Gouné [43] constructed a computational model of the precipitation kinetics of vanadium carbonitride in the Fe-V-C-N system steel that took into account the composition evolution of the precipitates with time. This model took advantage of the fast diffusion of nitrogen and carbon as compared to niobium to derive the composition, size and rate of formation of the precipitates during their nucleation.

For analysing the kinetics of dynamic precipitation in austenite of complex precipitates, Jung et al. [44] conducted a research on a steel microalloyed with niobium, titanium and vanadium during hot compression using electrical resistivity measurements, and found that electrical resistivity method could be applied to quantitatively evaluate the dynamic precipitation kinetics of complex precipitates in microalloyed steels. Based on the expression given by Dutta and Sellars [45], Medina et al. [46] constructed a model to study the strain-induced precipitation kinetics in microalloyed steels containing various combinations of carbon, nitrogen and precipitate-forming elements such as vanadium, niobium and titanium. This model took into account the influence of variables such as microalloying element percentages, strain, temperature, strain rate and grain size, and showed remarkable accuracy in predicting the PTT diagrams. In order to simulate precipitate particle growth resulting from collision and diffusion mechanisms in microalloyed steels, Xu and Thomas [47] developed computational particle-size-grouping (PSG) kinetic models based on population balances. This model was successfully applied to simulate the precipitated fraction of aluminium nitride and the size distribution of niobium carbide during isothermal aging process, and the predicted evolution of the particle size distribution were consistent with the results obtained by the classic nucleation, growth and coarsening theories.

3.3. Precipitate Solubility

Precipitates in microalloyed steels are normally carbides, nitrides or carbonitrides. Precipitation of these particles in austenite can restrict austenite grain growth and retard the occurrence of recrystallisation, and in ferrite can benefit the improving of strength. With the aim to make efficient use of microalloying elements and to understand well the influence of them on the mechanical properties and the hot working process of microalloyed steels, it is necessary to match their concentration with the processing conditions to control the dissolution and precipitation reactions of various kinds of precipitates. An understanding of the solubilities of precipitates is required.

Microalloying elements of primary interest include the strong precipitate formers vanadium, niobium and titanium. They form very stable carbides, nitrides or carbonitrides which may precipitate in austenite and/or ferrite depending on their relative solubilities and the processing history of the microalloyed steels. Aluminium is also of interest because of its role as an important deoxidiser and the precipitated AlN particles. Molybdenum in microalloyed steels may form complex carbonitrides with other alloying elements, and it has a role in the inhibiting of precipitates coarsening. In this section therefore, the solubilities of precipitates formed by these elements will be discussed. Tungsten carbide will also be addressed due to its growing academic and industrial interests in microalloyed steels.

The chemistry and stability of precipitates depend on the amount, type and number of alloying elements present, as well as the processing history of the steel. The stability of precipitate phases are normally expressed in terms of the solubility product of a metallic microalloying element [M] and a nonmetallic interstitial element [X] when they are both dissolved in austenite. The reaction between [M] and [X] to produce a precipitate MX can be expressed as:

$$[\mathrm{M}] + [\mathrm{X}] \leftrightarrow \mathrm{MX} \tag{3.1}$$

where [ ] signifies that the component is dissolved in austenite, X might be C or N, and MX a carbide or nitride phase. At equilibrium the following relation is satisfied:

$$k_{\mathrm{s}} = a_{[\mathrm{M}]} a_{[\mathrm{X}]} / a_{\mathrm{MX}} \tag{3.2}$$

where k_{s} is the solubility product, $a_{[\mathrm{M}]}$ and $a_{[\mathrm{X}]}$ are the activity coefficients of the dissolved [M] and [X], respectively, and a_{MX} is the activity of the carbide or nitride phase formed.

By assuming activities equal to the concentrations of the dissolved elements, an equating the activity of the phase formed to unity, k_{s} can be written as:

$$k_{\mathrm{s}} = [\mathrm{M}][\mathrm{X}] \tag{3.3}$$

If a precipitate is formed in the form of MX_y, Eq. (3.3) can be expressed as:

$$k_{\mathrm{s}} = [\mathrm{M}][\mathrm{X}]^y \tag{3.4}$$

where y is a constant average composition of X and defines stoichiometry. Then an Arrhenius relationship can be used to describe the temperature dependence of solubility product as:

$$\log_{10} k_{\mathrm{s}} = \log_{10}[\mathrm{M}][\mathrm{X}]^y = A - B/T \tag{3.5}$$

where A and B are experimentally determined constants for a given system, and T is the temperature in K. [M] and [X] are in weigh percentage or atomic percentage.

Table 3.1. Solubility products of precipitates in microalloyed steels

Compound	Metallic element	Nonmetallic element	A	B	Reference
VC	V	C	6.72	9500	[48]
$VC_{0.75}$, k_s=[V][C]$^{0.75}$	V	C	5.36	8000	[49]
$VC_{0.75}$	V	C	4.45	6500	[50]
VN	V	N	3.63	8700	[48]
VN	V	N	3.46	8330	[51]
VN	V	N	2.86	7700	[50]
VN	V	N	2.27	7070	[52]
NbC	Nb	C	4.37	9290	[53]
NbC	Nb	C	3.70	9100	[54]
NbC	Nb	C	3.42	7900	[48]
NbC	Nb	C	3.04	7290	[55]
NbC	Nb	C	2.26	6770	[51]
$NbC_{0.87}$, k_s=[Nb][C]$^{0.87}$	Nb	C	3.18	7700	[56]
$NbC_{0.87}$	Nb	C	3.11	7520	[55]
$NbC_{0.87}$	Nb	C	2.81	7020	[50]
NbN	Nb	N	4.04	10230	[54]
NbN	Nb	N	3.79	10150	[50]
NbN	Nb	N	3.70	10800	[54]
NbN	Nb	N	2.80	8500	[48]
$NbN_{0.87}$	Nb	N	2.86	7927	[57]
Nb(CN), k_s=[Nb][C+N]	Nb	C, N	1.54	5860	[55]
Nb(CN), k_s=[Nb][C+12/14N]	Nb	C, N	3.97	8800	[58]
Nb(CN), k_s=[Nb][C+12/14N]	Nb	C, N	2.26	6770	[51]
Nb(CN), k_s=[Nb][C]$^{0.83}$[N]$^{0.14}$	Nb	C, N	4.46	9800	[55]
Nb(CN), k_s=[Nb][C]$^{0.7}$[N]$^{0.2}$	Nb	C, N	4.12	9450	[50]
Nb(CN), k_s=[Nb][C]$^{0.24}$[N]$^{0.65}$	Nb	C, N	4.09	10500	[55]
TiC	Ti	C	5.33	10475	[59]
TiC	Ti	C	4.03	8720	[59]
TiC	Ti	C	2.75	7000	[51]
TiN	Ti	N	2.00	20790	[54]
TiN	Ti	N	6.75	19740	[54]
TiN	Ti	N	4.72	16192	[60]
TiN	Ti	N	5.40	15790	[50]
TiN	Ti	N	5.19	15490	[61]
TiN	Ti	N	3.93	15180	[62]
TiN	Ti	N	3.82	15020	[48]
TiN	Ti	N	0.32	8000	[59]
AlN	Al	N	1.80	7750	[49]
AlN	Al	N	1.48	7500	[63]
AlN	Al	N	1.95	7400	[54]
AlN	Al	N	1.03	6770	[51]
AlN	Al	N	0.725	6180	[64]
MoC	Mo	C	1.29	523	[65]
Mo_2C	Mo	C	3.04	2814	[65]
WC	W	C	5.00	6058	[65]
W_2C	W	C	3.61	3566	[65]

A summary of various investigations on the equilibrium solubility products of precipitates in austenite is given in terms of the *A* and *B* values in Table 3.1. It is seen in Table 3.1 that in some cases non-stoichiometric compounds in which a fraction of the carbon/nitrogen lattice sites are empty, e.g., $VC_{0.75}$, $NbC_{0.87}$ and $NbN_{0.87}$. The precise composition of a precipitate depends on the composition of the austenite matrix with which it is in equilibrium, as well as on the temperature. For example, the ratio of carbon to nitrogen in the niobium carbonitride Nb(CN) in equilibrium at a given temperature could be calculated for different amounts of carbon, nitrogen and niobium in microalloyed steels [50].

It has found that carbonitrides V(CN), Nb(CN) and Ti(CN) are completely intersoluable concerning the metallic elements in these precipitates, and that the entire spectrum of physical properties from one compound to the other is theoretically possible [66]. Also, because vanadium, niobium and titanium carbides, nitrides and carbonitrides have similar crystal structures, co-precipitation is very common in multi-microalloyed steels [67]. Therefore, complex or mixed precipitates may form easily during heat treatment and hot working of microalloyed steels.

The solubility products of precipitates that are of importance in microalloyed steels are compared in Figure 3.1 based on the values of *A* and *B* given in Table 3.1. It is clear that nitrides are more stable than carbides in austenite. The solubility differences between vanadium carbide and nitride, and between titanium carbide and nitride are significant, whilst the difference between niobium carbide and nitride is much less. The solubility difference between molybdenum carbide and tungsten carbide is very small, especially between Mo_2C and W_2C.

Titanium nitride is markedly more stable than any other kinds of carbides and nitrides. Vanadium carbide, on the other hand, is far more soluble in austenite compared with niobium and titanium carbides and nitrides, and as well as aluminium nitride.

Molybdenum and tungsten carbides have the lowest solubility among all the carbides and nitrides. Differences in solubility determine the way in which the use of these microalloying elements has developed in the control of the microstructure in microalloyed steels. It should be noted that the results in Figure 3.1 are obtained directly from the data presented in Table 3.1, and a careful check of the range of temperature in which the *A* and *B* values are valid should be made in relevant references before using these parameters.

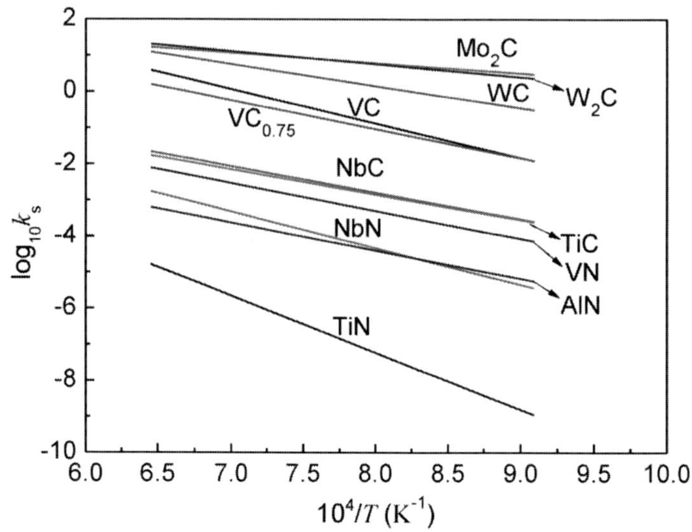

Figure 3.1. Solubility products of precipitates in austenite as function of temperature.

3.4. Nucleation and Growth of Precipitates

3.4.1. Classical Nucleation Theory

Precipitation of an individual particle involves its nucleation and growth. Classical nucleation theory, which is based on the change of Gibbs free energy (ΔG) associated with the formation of a precipitated particle in a supersaturated solid solution, is widely used to describe the nucleation process of precipitated phases in metal alloys. The free energy change associated with the nucleation process is dominated by chemical energy, strain energy and interfacial energy terms. In the case of a spherical embryo with a radius of r is created, ΔG is given by:

$$\Delta G = \frac{4}{3}\pi r^3 \Delta G_{chem} + \frac{4}{3}\pi r^3 \Delta G_{strain} + 4\pi r^2 E_i \tag{3.6}$$

where negative ΔG_{chem} is the chemical free energy change per unit volume, ΔG_{strain} is the strain energy per unit volume created from the formation of embryo, and E_i is the interfacial energy per unit area. At the right hand side of Eq. (3.6), the terms are respectively chemical energy, strain energy and interfacial energy. The relationship between ΔG and r is schematically illustrated in Figure 3.2. In the figure, ΔG^* is the activation energy for nucleation, and r^* is the critical radius at which the maximum free energy occurs. ΔG^* and r^* are also called the nucleation barrier and the critical radius for a stable nucleus respectively.

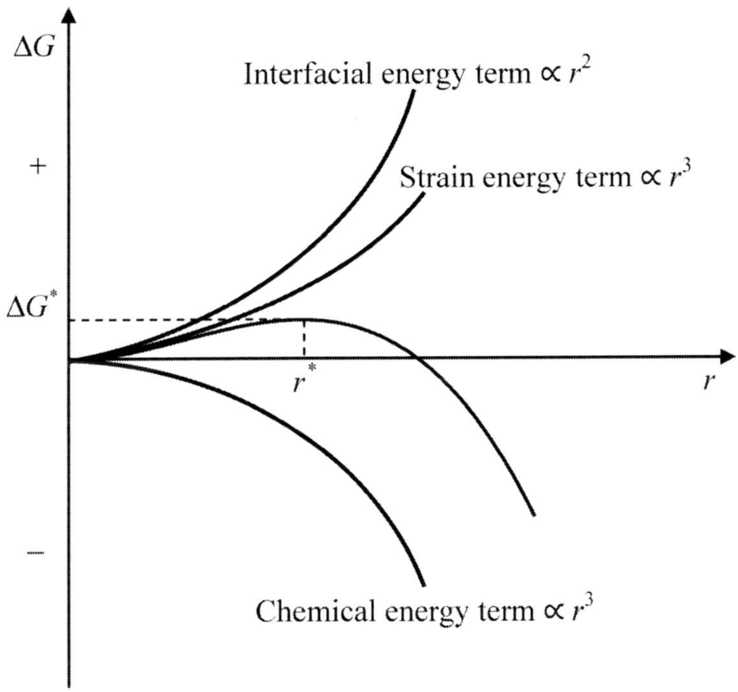

Figure 3.2. Schematic illustration of the free energy change associated with a nucleus formation as a function of radius r in the classical nucleation theory.

When the radius of a nucleus is greater than r^*, it will tend to grow rather than dissolve because its growth leads to a decreased free energy. When $r = r^*$, $d(\Delta G)/dr = 0$, ΔG^* is then given by:

$$\Delta G^{*} = \frac{16\pi}{3} \frac{E_i^3}{\left(\Delta G_{chem} + \Delta G_{strain}\right)^2} \tag{3.7}$$

The critical radius r^* is represented as:

$$r^{*} = -\frac{2E_i}{\Delta G_{chem} + \Delta G_{strain}} \tag{3.8}$$

The nucleation rate can be obtained by the following equation proposed by Becker and Döring [68] and Zeldovich [69] as:

$$I = N_0 Z \beta^{*} \exp\left(-\frac{\Delta G^{*}}{kT}\right) \tag{3.9}$$

where N_0 is the number of nucleation sites per unit volume, Z is the Zeldovich factor, β^* is the condensation rate of solute atoms in a cluster, k is the Boltzmann constant, and T is the absolute temperature. Z is obtained by:

$$Z = \frac{v_a^p}{2\pi r^{*}} \sqrt{\frac{E_i}{kT}} \tag{3.10}$$

where v_a^p is the mean atomic volume within precipitate.

β^* can be acquired by using the expression proposed by Russell [70] as:

$$\beta^{*} = \frac{4\pi r^{*2} D \chi}{a^4} \tag{3.11}$$

where D is the diffusion coefficient of solute atoms, χ is the matrix mean solute atom fraction, and a is the lattice parameter.

By introducing the incubation time for nucleation τ_i, the nucleation rate can be expressed as the derivative of the precipitate density N_n with respect to the time *t* of each phase [71]:

$$\left.\frac{dN_n}{dt}\right| = N_0 Z\beta^* \exp\left(-\frac{\Delta G^*}{kT}\right)\exp\left(-\frac{\tau_i}{t}\right) \tag{3.12}$$

where

$$\tau_i = \frac{2}{\pi\beta^* Z^2} \tag{3.13}$$

The nucleus size r_n^* can then be obtained by [72]:

$$r_n^* = r^* + \frac{1}{\sqrt{\pi Z}} \tag{3.14}$$

3.4.2. Growth of Precipitates

After nucleation, the stable nuclei are still embedded in a supersaturated matrix. The structure and composition at the interface between the matrix and a precipitated phase change discontinuously.

These discontinuities are maintained during the motion of this interface, and will constitute growth of the precipitated phase. In a binary system, the diffusion-controlled growth rate v_g of a spherical precipitate embedded in a supersaturated solid solution can be described as [73]:

$$v_g = \frac{dr}{dt} = \frac{\chi - \chi^i}{\varphi\chi^p - \chi^i}\frac{D}{r} \tag{3.15}$$

where χ is the mean solute mole fraction in the matrix, χ^i is the equilibrium solute mole fraction at the precipitate/matrix interface, χ^p is the mole composition of precipitate, and $\varphi = v_a^m / v_a^p$ is the ratio of matrix to precipitate atomic volumes (mean volume per atom). It should be noted that Eq. (3.15) is valid under the assumption of small supersaturation and local equilibrium. In a stoichiometric binary precipitate with composition of M_xX_y, radius of r and matrix/precipitate surface energy of E_i, the precipitate/matrix interface equilibrium mole fractions χ_M^i and χ_X^i can be modified as [74]:

$$\chi_M^i(r)^x \chi_X^i(r)^y = \chi_M^i(\infty)^x \chi_X^i(\infty)^y \exp\left(\frac{2E_i(x+y)v_a^p}{rkT}\right) \tag{3.16}$$

Then the growth rate can be expressed as:

$$v_g = \left.\frac{dr}{dt}\right|_{growth} = \frac{\chi - \chi^i(r)}{\varphi\chi^p - \chi^i(r)}\frac{D}{r} \tag{3.17}$$

3.5. Stability of Precipitates at Elevated Temperatures

In microalloyed steels, the austenite grain growth behaviour is closely related to the precipitates. Fine precipitate dispersions of the microalloying elements retard austenite growth, and the more stable the precipitates, the more effectively grain growth is retarded to higher temperatures. Therefore, in order to study the stability of precipitates such as carbides, nitrides and/or carbonitrides in microalloyed steels at elevated temperatures, the existing status of precipitates at different reheating temperatures should be identified.

Figure 3.3a shows the TEM micrograph of precipitates observed in the 0W steel austenitised at 950°C for 2 h followed by water quenching [67]. Figure 3.3b and c present the element analysis results obtained using energy dispersive X-ray spectroscopy (EDS) method on TEM [67]. As shown in Figure 3.3a, large numbers of precipitates exist at 950°C, and the precipitates can be classified into two types in shape, i.e., irregular shape and rectangle. Additional EDS analysis shows that both irregularly shaped (Figure 3.3b) and rectangular (Figure 3.3c) particles contain titanium, niobium and vanadium elements. Based on the compositions of the 0W steel given in Table 1.4 of

Chapter 1, the precipitates are probably titanium, niobium and vanadium carbides, nitrides or carbonitrides. Moreover, both spectra show high copper, silicon and oxygen peaks. The copper peaks are from the copper grid when preparing extraction replicas as well as thin foils mounted on copper grids for TEM observations [67]. The silicon and oxygen peaks may correspond to silica or silicic acid formed by oxidation and hydration of the silicon in the steel during the electrochemical extraction [75].

Figures 3.4, 3.5 and 3.6 show the morphology and compositions of precipitates in the 0W steel austenitised at 1100, 1150 and 1200°C for 2 h followed by water quenching, respectively [67]. At 1100°C, square and rectangle are found to be the two types of precipitates (Figure 3.4a). The square particle contains titanium, niobium and vanadium, while only titanium and niobium are found in rectangular particles, as seen in Figure 3.4b and c. When the austenitising temperature is elevated to 1150 and 1200°C, just rectangular precipitates, which only contain titanium and niobium, exist in the matrix, as shown in Figures 3.5 and 3.6.

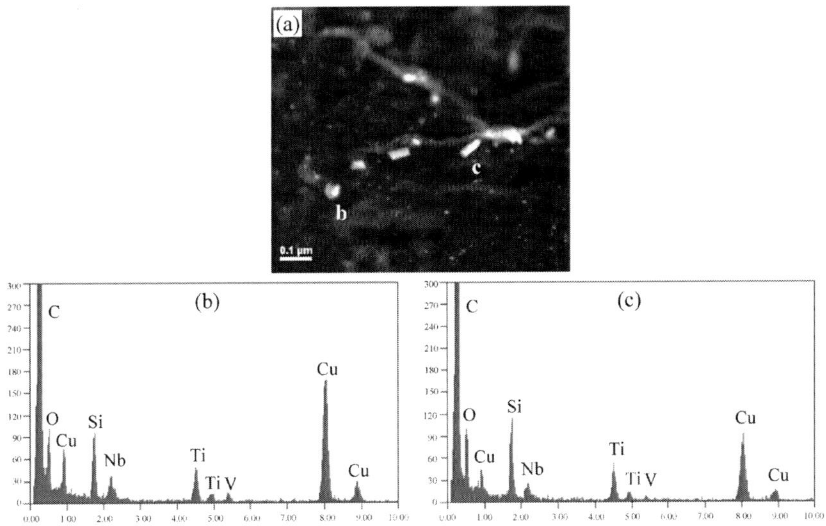

Figure 3.3. Morphology and compositions of precipitates in the 0W steel austenitised at 950°C for 2 h followed by water quenching: (a) Dark field image, (b) EDS of particle "b", and (c) EDS of particle "c" [67].

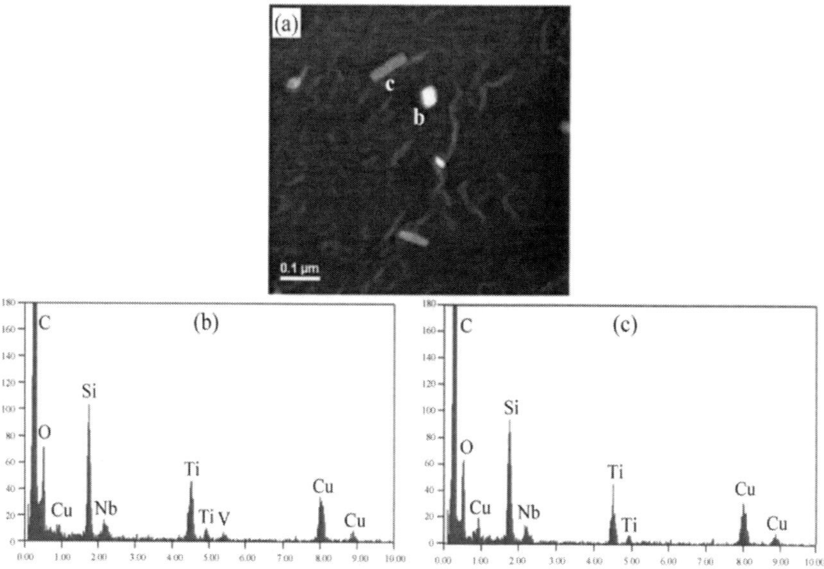

Figure 3.4. Morphology and compositions of precipitates in the 0W steel austenitised at 1100°C for 2 h followed by water quenching: (a) Dark field image, (b) EDS of particle "b", and (c) EDS of particle "c" [67].

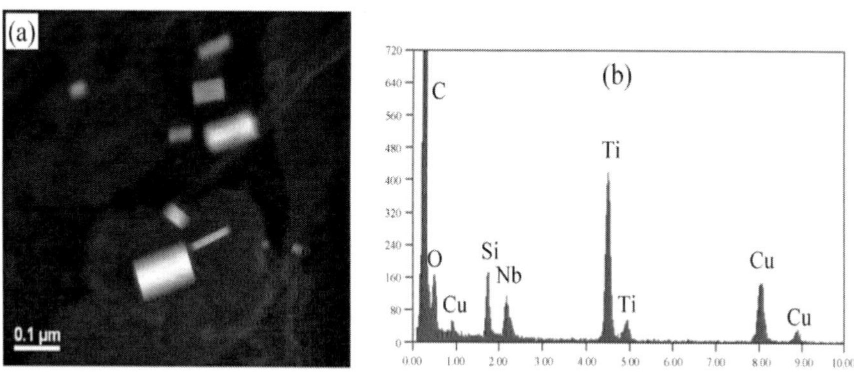

Figure 3.5. Morphology and compositions of precipitates in the 0W steel austenitised at 1150°C for 2 h followed by water quenching: (a) Dark field image, and (b) EDS of particle [67].

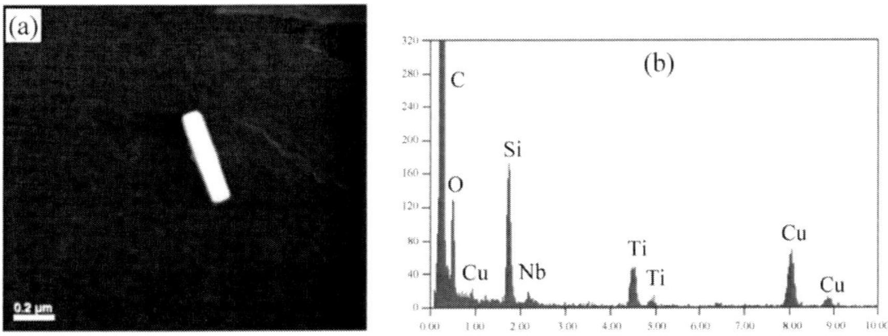

Figure 3.6. Morphology and compositions of precipitates in the 0W steel austenitised at 1200°C for 2 h followed by water quenching: (a) Dark field image, and (b) EDS of particle [67].

Table 3.2 shows the results of composition analysis of precipitates obtained from spectra [67]. As can be seen, increasing temperature from 950 to 1150°C leads to remarkable decrease in the content of niobium and vanadium in precipitates. Further increasing temperature to 1200°C causes leaner titanium content in the precipitate compared with that at 1150°C. Since titanium nitride TiN is the most stable and is known to have a dissolution temperature above the iron liquidus temperature, it is probable that TiN is the initial precipitate, and niobium or vanadium mostly forms on the outside layer of TiN to be the complex precipitate. In fact, because titanium, niobium and vanadium carbides, nitrides or carbonitrides have similar crystal structures, co-precipitation is very common in multi-microalloyed steels. Based on these results of the dissolution behaviour of precipitates, it is possible to predict the occurrence of coarsening of austenite grains at typical reheating temperatures, and therefore it will be valuable in the design of optimal heat treatment and hot working processes of microalloyed steels.

Table 3.2. Results of composition analysis of precipitates obtained from spectra [67]

Temperature (°C)	Precipitates	Composition ratio (at.%)
950	Irregularly shaped	Ti:Nb:V=1:0.91:0.12
950	Rectangular	Ti:Nb:V=1:0.42:0.02
1100	Square	Ti:Nb:V=1:0.32:0.02
1100	Rectangular	Ti:Nb=1:0.29
1150	Rectangular	Ti:Nb=1:0.28
1200	Rectangular	Ti:Nb=1:0.43

Note that the sizes of precipitates shown in Figures 3.3-3.6 do not exhibit significant change with the increase of reheating temperatures. It is because that the size of the precipitated particles in the steel is not uniform, and the selected TEM images are used only to present the typical morphologies of precipitates with different shapes. The average size of precipitates has actually been refined with the dissolution of niobium- and vanadium-containing particles from the outside layer of complex precipitates with an increase of the reheating temperatures, as shown in Figure 3.7 [76].

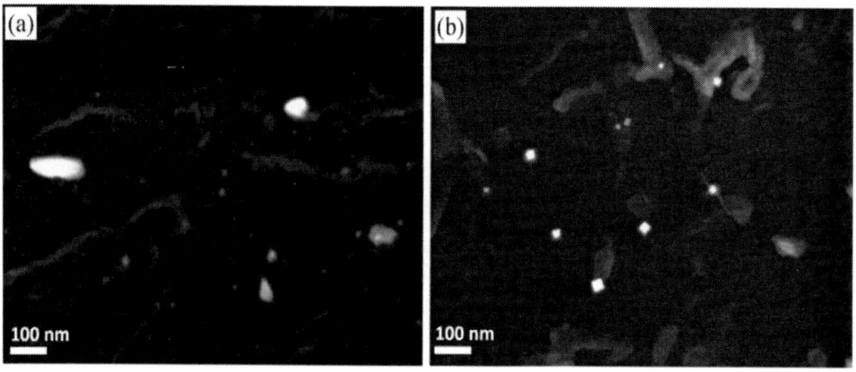

Figure 3.7. Precipitated particles observed in the 0W steel indicate that the average size of precipitates has been refined as the reheating temperature is raised: (a) 920°C + 1 h, and (b) 1230°C + 1 h [76].

3.6. Effect of Tungsten on Precipitate Characteristics

Figure 3.8 shows the morphology and compositions of precipitates observed in the as-forged 0W steel [77]. As illustrated in Figure 3.8a, six precipitates are found at the current magnification.

Two types of precipitates, i.e., rectangular and irregular precipitates, can be classified by shape. Additional EDS analysis shows that the rectangular particles contain titanium, niobium and vanadium, while only titanium and niobium are found in the irregular particles, as shown in Figure 3.8b and c.

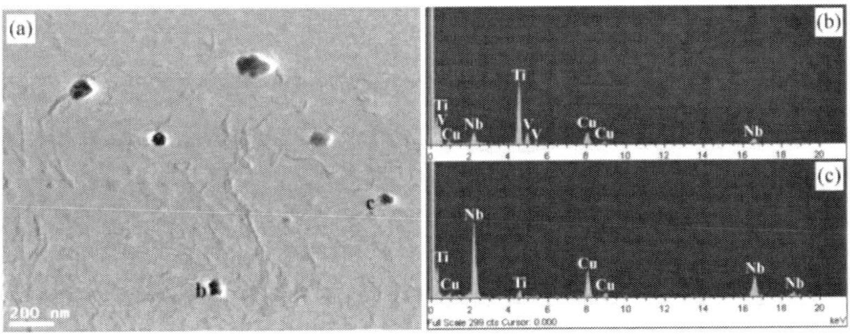

Figure 3.8. Morphology and compositions of precipitates in the as-forged 0W steel: (a) TEM bright field image, (b) EDS of particle "b", and (c) EDS of particle "c" [77].

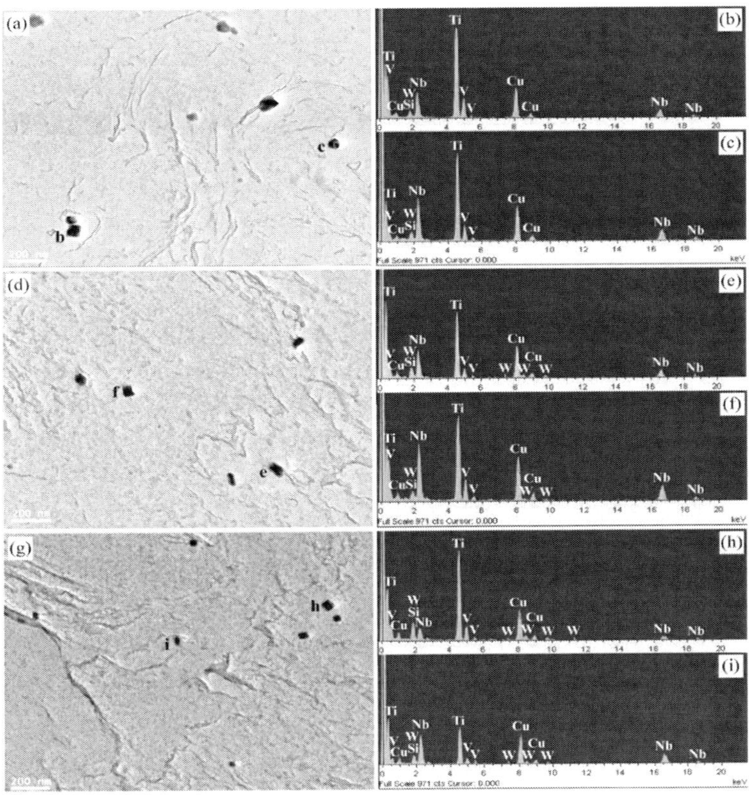

Figure 3.9. Morphology and compositions of precipitates in the as-forged (a,b,c) 0.1W, (d,e,f) 0.5W, and (g,h,i) 1W steels: (a,d,g) TEM bright field images, (b) EDS of particle "b", (c) EDS of particle "c", (e) EDS of particle "e", (f) EDS of particle "f", (h) EDS of particle "h", and (i) EDS of particle "i" [77].

Figure 3.9 shows the morphology and compositions of precipitates in the as-forged tungsten-containing microalloyed steels [77]. In the case of same magnification with that of the 0W steel, seven, five and seven precipitates are found in the 0.1W, 0.5W and 1W steels, respectively, as shown in Figure 3.9a, d and g. It is clear that the addition of W does not have a noticeable effect on the amount of precipitates.

EDS analysis shows that tungsten exists in both rectangular and irregular precipitates (Figure 3.9b, c, e, f, h and i).

Figure 3.10 shows the variation of average precipitate size with tungsten content [77]. It can be seen that the precipitate size decreases slowly after 0.1% tungsten addition, and then drops sharply when the tungsten content is increased from 0.1 to 0.5%, and again decreases slowly when the tungsten content is higher than 0.5%.

3.7. Concentration of Tungsten in Precipitates

The concentration of tungsten in different precipitates is presented in Table 3.3 [77]. It can be seen that the concentration of tungsten in rectangular precipitates is higher than that in irregular ones in the same steel. Also, it is obvious that the concentration of tungsten in both rectangular and irregular precipitates increases gradually with an increase of the tungsten content. According to these results, it is believed that increasing tungsten content results in higher tungsten concentration in precipitates, which subsequently retards the growth of such precipitates, as discussed in Section 3.5. Therefore, tungsten addition induces refined precipitates, which is expected to play an important role in the refinement of austenite grains. The effect of tungsten on the austenite grain size will be discussed in Chapter 4.

Table 3.3. Concentration of tungsten in different precipitates (wt.%) [77]

Steels	Rectangular precipitate	Irregular precipitate
0W	-	-
0.1W	3.9 ± 0.22	0.9 ± 0.36
0.5W	5.3 ± 0.64	2.8 ± 0.23
1W	12.8 ± 0.45	4.7 ± 0.82

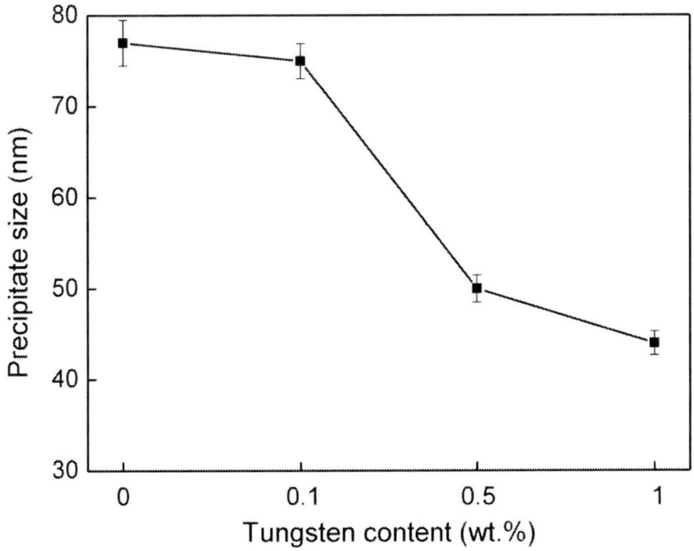

Figure 3.10. Variation of precipitate size with tungsten content [77].

3.8. SUMMARY

This chapter describes the roles of precipitates in both austenite and ferrite phases of microalloyed steels, and briefly reviews typical investigations on the precipitates in various microalloyed steels. The dissolution and precipitation reactions of various precipitates are described, and the solubility products of precipitates that can be commonly observed in microalloyed steels are discussed. The classical nucleation theory is introduced to help understanding the nucleation process of precipitated phases in microalloyed steels. The effects of reheating temperature on the status of precipitates are analysed. The shape, size and amount of precipitates in microalloyed forging steels with different tungsten additions are compared, and the concentration of tungsten in precipitates with different shapes is quantified.

The particles precipitated in microalloyed steel have a complex structure, in which TiN is probably the initial precipitate, and niobium or vanadium forms on the outside layer of TiN to be the complex precipitate. Increasing temperature from 950 to 1150°C leads to remarkable decrease in the content of niobium and vanadium in precipitates. Further increasing temperature to

1200°C causes leaner titanium content in the precipitate compared with that at 115°C.

The addition of tungsten does not have significant effect on the amount of precipitates in microalloyed steels. The precipitates in tungsten-containing microalloyed steels are all rich in tungsten, and the tungsten concentration in the precipitates increases with the increase of tungsten content. Tungsten addition in microalloyed steels has a beneficial effect on the retarding of precipitate growth.

REFERENCES

[1] C.L. Davis, M. Strangwood, Preliminary study of the inhomogeneous precipitate distribution in Nb-microalloyed plate steels, *Journal of Materials Science*, 37 (2002) 1083-1090.

[2] B. Dutta, C.M. Sellars, Strengthening of austenite by niobium during hot rolling of microalloyed steel, *Materials Science and Technology*, 2 (1986) 146-153.

[3] Q.Y. Sha, G.J. Huang, J. Guan, X.J. Ma, D.H. Li, A new route for identification of precipitates on austenite grain boundary in an Nb-V-Ti microalloyed steel, *Journal of Iron and Steel Research, International*, 18(8) (2011) 53-57.

[4] K.A. Pischow, A.K. Haraldsen, A.S. Korhonen, Scanning tunneling microscopy study of precipitates in a microalloyed steel, *Materials Characterization*, 31 (1993) 69-75.

[5] J.B. Wiskel, D.G. Ivey, H. Henein, The effects of finish rolling temperature and cooling interrupt conditions on precipitation in microalloyed steels using small angle neutron scattering, *Metallurgical and Materials Transactions B*, 39 (2008) 116-124.

[6] L. Rentería-Borja, E. Hurtado-Delgado, P. Garnica-González, I. Domínguez-López, A.L. García-García, Atomic force microscopy applied to the quantification of nano-precipitates in thermo-mechanically treated microalloyed steels, *Materials Characterization*, 69 (2012) 9-15.

[7] K. He, T.N. Baker, Zr-containing precipitates in a Ti-Nb microalloyed HSLA steel containing 0.016 wt.% Zr addition, *Materials Science and Engineering A*, 215 (1996) 57-66.

[8] J. Lu, J.B. Wiskel, O. Omotoso, H. Henein, D.G. Ivey, Matrix dissolution techniques applied to extract and quantify precipitates from a

microalloyed steel, *Metallurgical and Materials Transactions A*, 42 (2011) 1767-1784.

[9] J.S. Park, Y.K. Lee, Determination of Nb(C, N) dissolution temperature by electrical resistivity measurement in a low-carbon microalloyed steel, *Scripta Materialia*, 56 (2007) 225-228.

[10] S.G. Hong, K.B. Kang, C.G. Park, Stain-induced precipitation of NbC in Nb and Nb-Ti microalloyed HSLA steels, *Scripta Materialia*, 46 (2002) 163-168.

[11] K.J. Irvine, F.B. Pickering, T. Gladman, Grain-refined C-Mn steels, *Journal of the Iron and Steel Institute*, 205 (1967) 161-182.

[12] S.M. Medina, Determination of precipitation-time-temperature (PTT) diagrams for Nb, Ti or V micro-alloyed steels, *Journal of Materials Science,* 32 (1997) 1487-1492.

[13] S.Q. Yuan, G.L. Liang, Dissolving behaviour of second phase particles in Nb-Ti microalloyed steel, *Materials Letters*, 63 (2009) 2324-2326.

[14] S.G. Hong, H.J. Jun, K.B. Kang, C.G. Park, Evolution of precipitates in the Nb-Ti-V microalloyed steels during reheating, *Scripta Materialia*, 48 (2003) 1201-1206.

[15] K. Narita, Physical chemistry of the groups IVa (Ti, Zr), Va (V, Nb, Ta) and the rare earth elements in steel, *Transactions of the Iron and Steel Institute of Japan*, 15 (1975) 145-152.

[16] E.T. Turkdogan, Causes and effects of nitride and carbonitride precipitation during continuous casting, *Iron and Steelmaker*, 16 (1989) 61-75.

[17] S. Vervynckt, K. Verbeken, P. Thibaux, Y. Houbaert, Recrystallization-precipitation interaction during austenite hot deformation of a Nb microalloyed steel, *Materials Science and Engineering A*, 528 (2011) 5519-5528.

[18] X.J. Zhuo, D.H. Woo, X.H. Wang, H.G. Lee, Formation and thermal stability of large precipitates and oxides in titanium and niobium microalloyed steel, *Journal of Iron and Steel Research, International*, 15(3) (2008) 70-77.

[19] S. Gündüz, R.C. Cochrane, Influence of cooling rate and tempering on precipitation and hardness of vanadium microalloyed steel, *Materials and Design*, 26 (2005) 486-492.

[20] H.R. Wang, W. Wang, Precipitation of complex carbonitrides in a Nb-Ti microalloyed plate steel, *Journal of Materials Science*, 44 (2009) 591-600.

[21] F. Ma, G. Wen, P. Tang, G. Xu, F. Mei, W. Wang, Effect of cooling rate on the precipitation behavior of carbonitride in microalloyed steel, *Metallurgical and Materials Transactions B*, 42 (2011) 81-86.
[22] Q.Y. Sha, G.Y. Li, L.F. Qiao, P.Y. Yan, Effect of cooling rate and coiling temperature on precipitate in ferrite of a Nb-V-Ti microalloyed strip steel, *Proceedings of Sino-Swedish Structural Materials Symposium*, 2007, pp. 316-319.
[23] M. Longauerová, P. Bekeč, M. Fedorová, S. Longauer, Influence of casting rate on precipitation of particles based on micro-alloying elements in slab surface zone, *Acta Metallurgical Slovaca*, 17 (2011) 177-185.
[24] K. Xu, B.G. Thomas, R. O'Malley, Equilibrium model of prediction in microalloyed steels, *Metallurgical and Materials Transactions A*, 42 (2011) 524-539.
[25] M. Arribas, B. López, J.M. Rodriguez-Ibabe, Additional grain refinement in recrystallization controlled rolling of Ti-microalloyed steels processed by near-net-shape casing technology, *Materials Science and Engineering A*, 485 (2008) 383-394.
[26] J. Lu, O. Omotoso, J.B. Wiskel, D.G. Ivey, Strengthening mechanisms and their relative contributions to the yield strength of microalloyed steels, *Metallurgical and Materials Transactions A*, 43 (2012) 3043-3061.
[27] R.D.K. Misra, Z. Jia, R. O'Malley, S.J. Jansto, Precipitation behavior during thin slab thermomechanical processing and isothermal aging of copper-bearing niobium-microalloyed high strength structural steels: The effect on mechanical properties, *Materials Science and Engineering A*, 528 (2011) 8772-8780.
[28] H. Wu, L. Du, Z. Ai, X. Liu, Static recrystallization and precipitation behavior of a weathering steel microalloyed with vanadium, *Journal of Materials Science and Technology*, 29(12) (2013) 1197-1203.
[29] X. Li, L. Zhao, X. Wang, Y. Zhao, Precipitation and hetero-nucleation effect of V(C,N) in V-microalloyed steel, *Journal of Wuhan University of Technology-Materials Science Edition*, 23(6) (2008) 844-849.
[30] E.V. Pereloma, A.D. Kostryzhev, A. AlShahrani, C. Zhu, J.M. Cairney, C.R. Killmore, S.P. Ringer, Effect of austenite deformation temperature on Nb clustering and precipitation in microalloyed steel, *Scripta Materialia*, 75 (2014) 74-77.
[31] C. Wagner, *Thermodynamics of Alloys*, Addison-Wesley Publishing Company, Cambridge, MA, 1952.

[32] I. I. Gorbachev, V. V. Popov, Analysis of the solubility of carbides, nitrides, and carbonitrides in steels using methods of computer thermodynamics: IV. Solubility of carbides, nitrides, and carbonitrides in the Fe-Nb-C, Fe-Nb-N, and Fe-Nb-C-N systems, *The Physics of Metals and Metallography*, 110 (2010) 52-61.

[33] Y. Luo, J. zhang, C. Xiao, Thermodynamic calculation and experimental investigation of second phase particles in low carbon Nb microalloyed steels, *Steel Research International*, 83 (2012) 238-243.

[34] S. Xiang, G. Liu, Y. Li, C. Li, A. Wang, Thermodynamics research on precipitates in low carbon Nb-microalloyed steels produced by compact strip production, *Journal of Materials Science and Technology*, 23 (2007) 267-272.

[35] J.G. Speer, J.R. Michael, S.S. Hansen, Carbonitride precipitation in niobium/vanadium microalloyed steels, *Metallurgical Transactions A*, 18 (1987) 211-222.

[36] W.J. Liu, J.J. Jonas, Calculation of the Ti(CyN1-y)-Ti4C2S2-MnS-austenite equilibrium in Ti-bearing steels, *Metallurgical Transactions A*, 20 (1989) 1361-1374.

[37] N. Gao, T.N. Baker, Influence of AlN precipitation on thermodynamic parameters in Al-V-N microalloyed steels, *ISIJ International*, 37 (1997) 596-604.

[38] H. Adrian, Thermodynamic model for precipitation of carbonitrides in high strength low alloy steels containing up to three microalloying elements with or without additions of aluminium, *Materials Science and Technology*, 8 (1992) 406-420.

[39] P.R. Rios, Effect of aluminium on the equilibrium precipitation of a multicomponent f.c.c. carbonitride in microalloyed steels, *Materials Science and Engineering*, A156 (1992) L5-L8.

[40] Y. Zeng, W. Wang, Computer simulation of carbonitride precipitation during deformation in Nb-Ti microalloyed steels, *Journal of Materials Science*, 43 (2008) 874-882.

[41] J.S. Park, Y.S. Ha, S.J. Lee, Y.K. Lee, Dissolution and precipitation kinetics of Nb(C,N) in austenite of a low-carbon Nb-microalloyed steel, *Metallurgical and Materials Transactions A*, 40 (2009) 560-568.

[42] Z. Wang, X. Sun, Z. Yang, Q. Yong, C. Zhang, Z. Li, Y. Weng, Effect of Mn concentration on the kinetics of strain induced precipitation in Ti microalloyed steels, *Materials Science and Engineering A*, 561 (2013) 212-219.

[43] P. Maugis, M. Gouné, Kinetics of vanadium carbonitride precipitation in steel: A computer model, *Acta Materialia*, 53 (2005) 3359-3367.

[44] J.G. Jung, J.H. Bae, Y.K. Lee, Quantitative evaluation of dynamic precipitation kinetics in a complex Nb-Ti-V microalloyed steel using electrical resistivity measurements, *Metals and Materials International*, 19 (2013) 1159-1162.

[45] B. Dutta, C.M. Sellars, Effect of composition and process variables on Nb(C, N) precipitation in niobium microalloyed austenite, *Materials Science and Technology*, 3 (1987) 197-206.

[46] S.F. Medina, A. Quispe, M. Gomez, Model for strain-induced precipitation kinetics in microalloyed steels, *Metallurgical and Materials Transactions A*, 45 (2014) 1524-1539.

[47] K. Xu, B.G. Thomas, Particle-size-grouping model of precipitation kinetics in microalloyed steels, *Metallurgical and Materials Transactions A*, 43 (2012) 1079-1096.

[48] K. Narita, Physical Chemistry of the group IVa (Ti, Zr), Va (V, Nb, Ta) and rare elements in steel, *Transactions ISIJ*, 15 (1975) 145-152.

[49] M.F. Ashby, K.E. Easterling, A first report on diagrams for grain growth in welds, *Acta Metallurgica*, 30 (1982) 1969-1978.

[50] D. Llewellyn, R. Hudd, *Steels: Metallurgy and Applications*, Butterworth-Heinemann, 1998.

[51] K. Irvine, F. Pickering, T. Gladman, Grain-refined C-Mn steels, *Journal of the Iron and Steel Institute*, 205(2) (1967) 161-182.

[52] R.W. Fountain, J. Chipman, Solubility and precipitation of vanadium nitride in alpha and gamma iron, *Transactions of the Metallurgical Society of AIME,* 212 (1958) 737-739.

[53] T.H. Johansen, N. Christensen, B Augland, The solubility of niobium carbide in gamma iron, *Transactions of the Metallurgical Society of AIME*, 239 (1967) 1651-1654.

[54] J. Strid, K.E. Easterling, On the chemistry and stability of complex carbides and nitrides in microalloyed steels, *Acta Metallurgica*, 33 (1985) 2057-2074.

[55] H. Nordberg, B. Aronsson, Solubility of niobium carbide in austenite, *Journal of the Iron and Steel Institute*, 206(12) (1968) 1263-1266.

[56] T. Mori, M. Tokizane, K. Yamaguchi, E. Sunami, Y. Nakazima, Thermodynamic properties of niobium carbides and nitrides in steels, *Tetsu-to-Hagane*, 54 (1968) 763-776.

[57] R. C. Sharma, V. K. Lakshmanan, J. S. Kirkaldy, Solubility of niobium carbide and niobium carbonitride in alloyed austenite and ferrite, *Metallurgical Transactions A*, 15A (1984) 545-553.
[58] M.L. Santella, Grain Growth and High-Temperature Hot Rolling Behavior of Low-Alloy Steel Austenite, University of Pittsburgh, 1981.
[59] K. Balasubramanian, *Thermodynamics of Microalloyed Austenites and Nonstoichiometric Carbides and Nitrides*, McMaster University, 1988.
[60] H. Sawamura, T. Mori, *Thermodynamic investigation of non-metallic inclusions in steel (II)*, Tetsu-to-Hagane, 43(1) (1957) 31-38.
[61] J. Kunze, Solubility product of titanium nitride in γ-iron, *Metal Science*, 16 (1982) 217-218.
[62] K. Narita, Elements present in minute amounts in steel and iron. III. Titanium and zirconium nitride, *Journal of the Chemical Society of Japan*, 77 (1956) 1536-1539.
[63] T Gladman, F.B. Pickering, Grain coarsening of austenite, *Journal of the Iron and Steel Institute*, 205 (1967) 653-664.
[64] L.A. Erasmus, G.I Meck, Effect of aluminum additions on forgeability, austenite grain coarsening temperature, and impact properties of steel, *Journal of the Iron and Steel Institute*, 202 (1964) 32-41.
[65] E.J. Pavlina, J.G. Speer, C.J. Van Tyne, Equilibrium solubility products of molybdenum carbide and tungsten carbide in iron, *Scripta Materialia*, 66 (2012) 243-246.
[66] D.C. Houghton, G.C. Weatherly, J.D. Embury, Microchemistry of carbonitrides in the HAZ of HSLA steels, In: *Proceeding of the International Conference on Advances in the Physical Metallurgy and Applications of Steels*, Liverpool, UK, 1981. The Metals Society, London, 1982, pp. 136-146.
[67] J. Zhao, J.H. Lee, Y.W. Kim, Z. Jiang, C.S. Lee, Enhancing mechanical properties of a low-carbon microalloyed cast steel by controlled heat treatment, *Materials Science and Engineering A*, 559 (2013) 427-435.
[68] R. Becker, W. Döring, Kinetische behandlung der keimbildung in übersättigten dämpfen, *Annalen der Physik*, 416 (1935) 719-752.
[69] Y.B. Zeldovich, On the theory of new phase formation: cavitation, *Acta Physicochim USSR*, 18 (1943) 1-22.
[70] K.C. Russell, *Phase transformations, American Society for Metals*, Ohio, 1968.
[71] R. Wagner, R. Kampmann, Homogeneous second phase precipitation, in: P. Haasen (Ed.), *Materials Science and Technology: A Comprehensive Treatment*, Wiley-VCH, Weinheim, 1991.

[72] M. Perez, M. Dumont, D. Acevedo-Reyes, Implementation of classical nucleation and growth theories for precipitation, *Acta Materialia*, 56 (2008) 2119-2132.

[73] C. Zener, Theory of growth of spherical precipitates from solid solution, *Journal of Applied Physics*, 20 (1949) 950-953.

[74] M. Perez, Gibbs-Thomson effect in phase transformations, *Scripta Materialia,* 52 (2005) 709-712.

[75] R.W. Gurry, J. Christakos, C.D. Stricker, Electrolytic extraction of carbides from carbon steel, *Transactions ASM*, 50 (1958) 105-128.

[76] J. Zhao, Z. Jiang, C.S. Lee, Enhancing impact fracture toughness and tensile properties of a microalloyed cast steel by hot forging and post-forging heat treatment processes, *Materials and Design*, 47 (2013) 227-233.

[77] J. Zhao, T. Lee, J.H. Lee, Z. Jiang, C.S. Lee, Effects of tungsten addition on the microstructure and mechanical properties of microalloyed forging steels, *Metallurgical and Materials Transactions A*, 44A (2013) 3511-3523.

Chapter 4

EFFECT OF TUNGSTEN ON MICROSTRUCTURAL EVOLUTION

ABSTRACT

This chapter describes the microstructural evolution of microalloyed steels after tungsten addition. Austenite grain size, microstructural constituent and feature, and interlamellar spacing of pearlite are all important characteristics of the microstructural evolution of microalloyed steels. These features are described and related to the major morphologies. Etching methods for revealing prior austenite grain boundaries are first introduced. The dependences of austenite grain size on austenitising temperature and holding time are presented. The difference in austenite grain size between the tungsten-free and tungsten-containing microalloyed steels is compared. Then the microstructures of microalloyed forging steels with different tungsten additions are analysed, and then a simple and universal measurement method is introduced to determine the average true interlamellar spacing of pearlite of microalloyed steels. Finally, the mechanism of tungsten-induced microstructural evolution is discussed.

4.1. MICROSTRUCTURE OF MICROALLOYED STEELS

As described in Chapter 2, austenite is the parent phase of all the microstructures including pearlite, pro-eutectoid phases, bainite, martensite, and various ferritic microstructures of microalloyed steels. Carbon has a high solubility in the octahedral interstitial sites of close-packed arrays of iron atoms in austenite which has a FCC crystal structure. Such high solubility of carbon in austenite causes cementite to dissolve and the carbon concentrated in cementite to go into solution in austenite. As a result, austenite is expected to have a very high hot ductility and is readily hot worked by forging and rolling to smaller sections and complex shapes due to the absence of high densities of second phases such as cementite and other carbides [1]. Austenite is a major contributor to the cost-effective manufacture of steel structures especially when a heavy section must be converted into smaller sections and shapes. Generally, the austenite of a microalloyed steel can be transformed to all of the aforementioned microstructures depending on the chemical compositions and cooling rate. Some of the microstructures might serve in-process functions and some might serve for final applications. In some cases, retained austenite may exist in the final microstructure due to the variations of alloying elements and cooling rate after hot working. The microstructure of microalloyed steels can be modified by changing the chemical compositions and hot working conditions according to the requirement of mechanical properties for specified applications.

4.2. Effect of Tungsten on Austenite

4.2.1. Determination of Austenite Grain Size

Austenite grain size has a prominent influence on the characteristics of microalloyed steels such as hardenability, mechanical properties, and the microstructures of austenite transformation products. Therefore, determination of austenite grain size of microalloyed steels becomes of the greatest importance in metallurgical studies. Austenite, however, is no longer present in most transformed microstructures at room temperature, and even if some is retained as in medium- and high-carbon martensitic microstructures, the retained austenite does not define the grain size of the parent austenite. Thus, the austenite grain size of a steel is usually referred to as the prior austenite grain size.

To determine the prior austenite grain size of microalloyed steels, two distinct stages are involved: revealing the prior austenite grain boundaries and measuring the grain sizes. In medium- and high-carbon steels, delineation of thin grain boundary networks of pro-eutectoid ferrite or pro-eutectoid cementite within largely pearlitic microstructures that are produced after hot working or normalising is very effective in revealing prior austenite grain boundaries. In microalloyed steels where the carbon content is low, the equiaxed ferrite grains grow across austenite grain boundaries and make up most of the microstructure, delineation of austenite grain boundaries by ferrite is therefore not effective. Quenching from austenite phase field to produce martensite that marks the extent of austenite grains is a very effective method and has been used extensively to reveal prior austenite grain boundaries in microalloyed and other low-carbon steels.

For revealing prior austenite grain boundaries, several metallographic etching techniques have been used. Table 4.1 presents some of the methods used to reveal prior austenite grain boundaries in microalloyed steels. Among these methods, some of them are adaptable to general use and others are designed for a specific set of conditions.

For a microalloyed steel quenched from the austenite phase field, nital etching cannot be used to show the prior austenite grain boundaries, but it tends to reveal primarily the details of martensitic microstructure. An example of etching to reveal the martensitic microstructure and prior austenite grain boundaries is shown in Figure 4.1. Nital etching does not reveal the austenite grain boundaries, as shown in Figure 4.1a. The austenite grain boundaries shown in Figure 4.1b are revealed by etching the quenched specimen in a solution of "5% picric acid (5 g picric acid and 285 ml distilled water) + 10% $FeCl_3$ (10 g $FeCl_3$ and 90 ml distilled water) + 10% dodecyl benzene sulfonic acid sodium salt (10 g dodecyl benzene sulfonic acid sodium salt and 90 ml distilled water)" for 10 min at room temperature. The prior austenite grain boundaries have been well revealed by etching the quenched microalloyed steel in this solution.

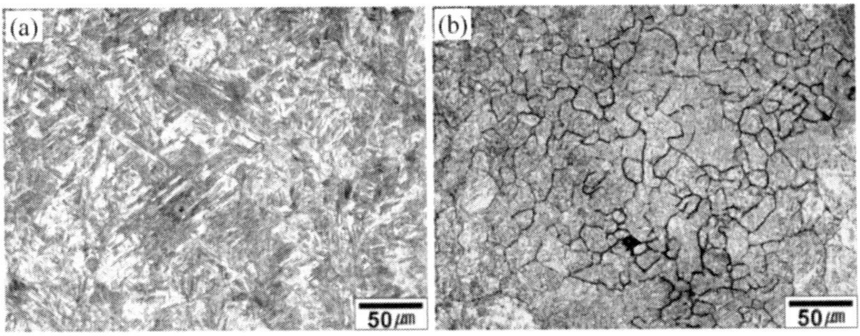

Figure 4.1. Microstructure of the 0W steel austenitised at 1100°C for 2 h followed by water quenching: (a) etched to show martensite with 2% nital, and (b) etched to reveal prior austenite grain boundaries with the "method 10" given in Table 4.1.

Table 4.1. Etching methods for revealing prior austenite grain boundaries

Method	Etchant	Operating condition	Steel type	Carbon content	Reference
1	Saturated picric solution containing a few drops detergent	80°C, 7 min	Nb-Ti-V microalloyed steel	0.04 wt.% C	[2]
2	Saturated picric acid containing 2-4 drops of dimethylbenzene	70-80°C	High Nb microalloyed steel	0.06 wt.% C	[3]
3	4 g sodium dodecylbenzene sulfonic acid + 100 ml saturated aqueous picric acid + 100 ml distilled water + a few drops of Triton X-100 surface active reagent	60-70°C, 60-180 s	Nb-Ti microalloyed steel	0.06 wt.%	[4]
4	0.3% hydrochloric acid + 3% dodecylbenzenesulfonic acid + 96.7% saturated picric acid	70°C	Nb-Ti-V microalloyed steel	0.08 wt.% C	[5]
5	Saturated solution of aqueous picric acid, hydrochloric acid and detergent	68°C, 5 min	Nb-Ti microalloyed steel	0.081 wt.% C	[6]
6	1 g $C_2H_2O_4$ + 100 ml H_2O + 5 ml H_2O_2	-	Nb-Ti-V-Zr microalloyed steel	0.09 wt.% C	[7]
7	Saturated solution of aqueous picric acid	60°C	Nb microalloyed steel	0.1 wt.% C	[8]
8	100 ml saturated aqueous picric acid + 10 ml teepol	60°C, 4-6 min	Nb-Ti microalloyed steel	0.11 wt.% C	[9]
9	4 g picric acid + 100 ml ethanol + 1 ml HCl	-	Ti microalloyed steel	0.15 wt.% C	[10]
10	Picric acid + $FeCl_3$ + SDS (dodecyl benzene sulfonic acid sodium salt)	Room temperature	Nb-Ti-V-W microalloyed steels	0.17 wt.% C	[11]

Prior austenite grain size can be predicted by comparison to standards. The comparison method involves comparison of the grain structure to a series of chart ratings. The ASTM grain size number, *G*, is defined as:

$$n_{si} = 2^{G-1} \tag{4.1}$$

where n_{si} is the number of grains per square inch at 100× magnification. To obtain the number per square millimetre at 1×, multiply by 15.5. The relationship can also be expressed as:

$$G = \frac{\log_{10} n_{si}}{\log_{10} 2} + 1 \tag{4.2}$$

The comparison method is commonly used in the prediction of grain size. However, chart ratings of grain size are biased, and the grain size is somewhat coarser (0.5 to 1 *G* number lower) than it actually is. The repeatability and reproducibility of comparison chart ratings are generally ±1 grain size number.

Prior austenite grain size can also be measured by planimetric method or intercept method. In the planimetric method, an actual count of the number of grains is made within a known area. The number of grains per unit area, n_a, is used to determine the ASTM grain size number, *G*. The precision of the method is a function of the number of grains counted. A precision of ±0.25 grain size units can be attained with a reasonable amount of effort. Results are free of bias and repeatability and reproducibility are less than ±0.5 grain size units. The magnification of the image should be known and kept to minimum possible. This method is slow when done manually because the grains must be marked when counted to obtain an accurate count.

The intercept method involves an actual count of the number of grains intercepted, *N*, by a test line (circular or straight) or the number of grain boundary interceptions, *P*, with a test line (circle or straight).

In case of a circle line, *N* or *P* is divided by the true line length, L_t ($L_t = \pi D / \text{magnificat ion}$, where *D* is the diameter of the circle), to determine the number of interceptions per unit length N_L or P_L by: $N_L = N / L_t$ or $P_L = P / L_t$. The reciprocal of N_L or P_L gives the average lineal intercept length, $\bar{l}$: $\bar{l} = 1/N_L$ or $\bar{l} = 1/P_L$. $\bar{l}$ is used to determine the ASTM grain size number, G, by:

$$G = -3.2877 - 6.6439 \log_{10} \bar{l} \tag{4.3}$$

where $\bar{l}$ is in mm. The precision of the method is a function of the number of intercepts or intersections counted. A precision of better than ±0.25 grain size units can be attained with a reasonable amount of effort. Results are free of bias; repeatability and reproducibility are less than ±0.5 grain size units. The intercept method is more efficient than the planimetric method yielding acceptable measurement precision in much less time. In practice, we should repeat the measurements on a number of filed in order to obtain a good estimate of the grain size. Refer to ASTM E 112 for a detailed description of these methods.

4.2.2. Austenite Grain Size of Tungsten-Free Microalloyed Steel

Figure 4.2a and b present the SEM images of prior austenite grains of the water-quenched tungsten-free (0W) steel austenitised at 950°C and 1200°C for 2 h, respectively. It can be seen that increment of austenitising temperature from 950 to 1200°C induces significant increase in the average prior austenite grain size.

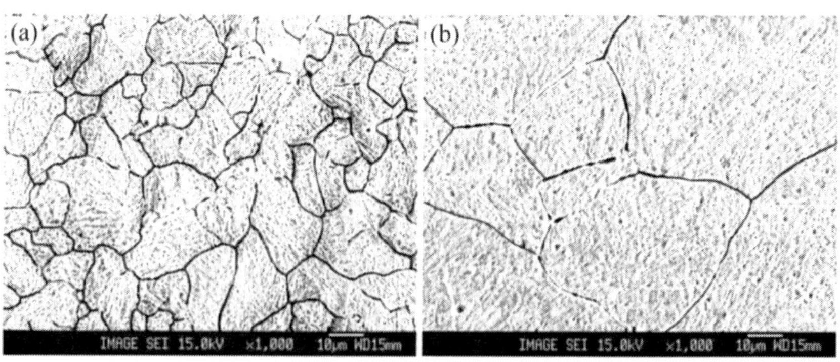

Figure 4.2. Prior austenite grains of the 0W steel austenitised at (a) 950°C and (b) 1200°C for 2 h followed by water quenching.

Figure 4.3 shows the dependence of prior austenite grain size on austenitising temperature [12]. As can be seen, the prior austenite grain size slightly increases as the temperature increases from 950 to 1100°C. When the temperature is higher than 1100°C, sharp increase in prior austenite grain size is observed.

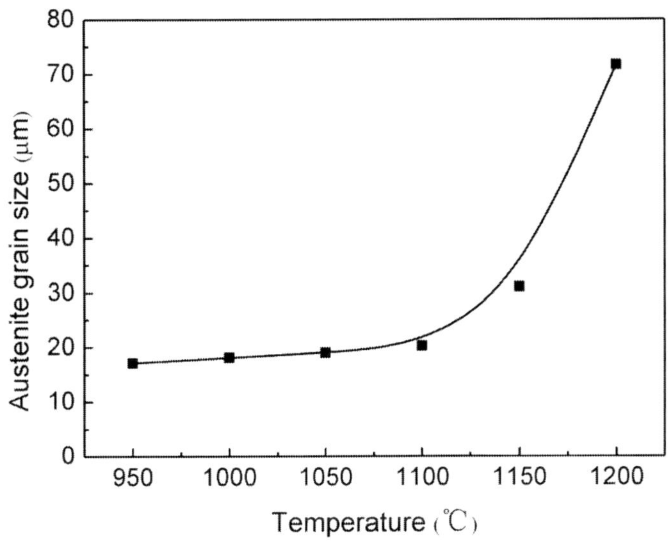

Figure 4.3. Effect of temperature on the prior austenite grain size of the water-quenched 0W steel with holding time of 2 h [12].

Combined with the results presented in Figures 3.3-3.6 and Table 3.2 of Chapter 3, it can be deduced that the dissolution of niobium- and vanadium-containing precipitates has no significant effect on the austenite grain growth when the temperature is lower than 1100°C. However, dissolution of vanadium-containing precipitates plays a key role in the abnormal growth of austenite grains when the temperature is increased from 1100 to 1150°C. When the austenitising temperature is increased to 1200°C, significant reduction of the amount of precipitates will be caused. Therefore, further growth in austenite grains at 1200°C should be attributed to the dissolution of titanium-containing particles and the reduction of the amount of precipitates.

Figure 4.4 shows the prior austenite grains of the water-quenched 0W steel austenitised at 1100°C for both 0.5 and 5 h. No significant change in the prior austenite grain size occurs when the holding time is increased from 0.5 to

5 h. Holding time shows little influence on the prior austenite grain size, as indicated in Figure 4.5 [12].

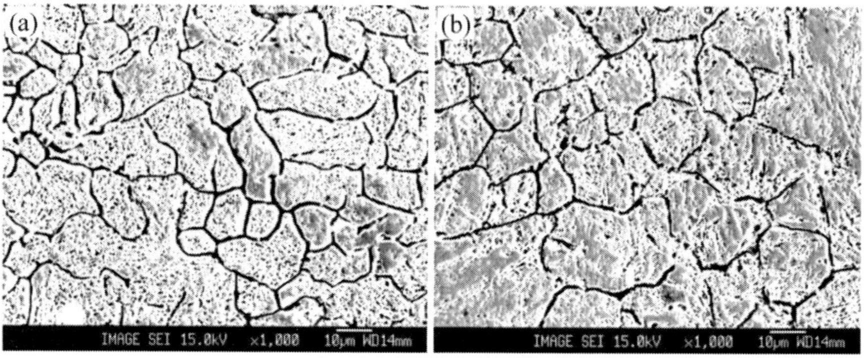

Figure 4.4. Prior austenite grains of the 0W steel austenitised at 1100°C for (a) 0.5 h and (b) 5 h followed by water quenching.

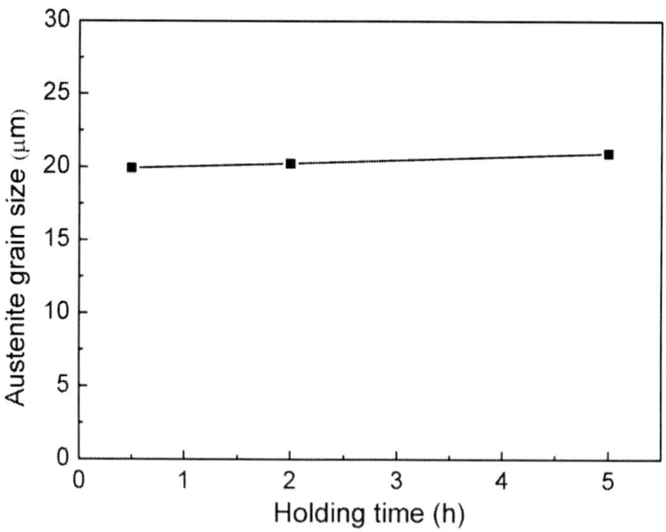

Figure 4.5. Effect of holding time on the prior austenite grain size of the water-quenched 0W steel with austenitising temperature of 1100°C.

4.2.3. Effect of Tungsten on Austenite Grain Size

As shown in Figure 4.6 [13], coarse-grained austenite exists in the 0W steel austenitised at 1230°C for 1 h followed by water quenching. After determination by using the circular intercept method according to ASTM E112, the average prior austenite grain size is obtained to be 158.6 μm.

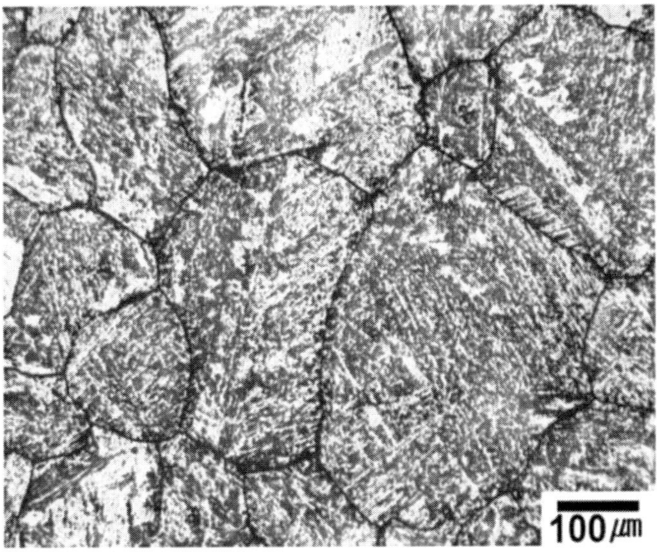

Figure 4.6. Prior austenite grains of the 0W steel austenitised at 1230°C for 1 h followed by water quenching [13].

The prior austenite grains of the water-quenched tungsten-containing steels austenitised at 1230°C for 1 h are shown in Figure 4.7 [13]. The addition of tungsten has greatly changed the prior austenite grain size, and increasing the tungsten content leads to finer prior austenite grain size, as shown in Figure 4.7a through d. It can therefore be concluded that tungsten has a positive effect on retarding the growth of austenite grains. Figure 4.8 shows the dependence of average prior austenite grain size on tungsten content [13].

As can be seen, the prior austenite grain size decreases slightly as the tungsten content increases from 0 to 0.1%, then drops sharply from 0.1% to 0.5%, and finally decreases gradually when the tungsten content is greater than 0.5%.

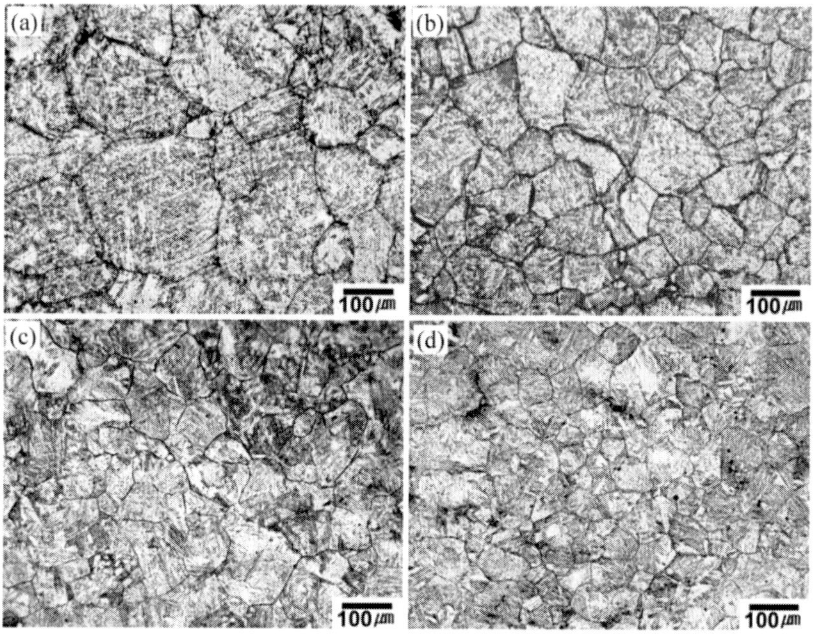

Figure 4.7. Prior austenite grains of the tungsten-containing steels austenitised at 1230°C for 1 h followed by water quenching: (a) 0.1W, (b) 0.5W, (c) 1W, and (d) 2W [13].

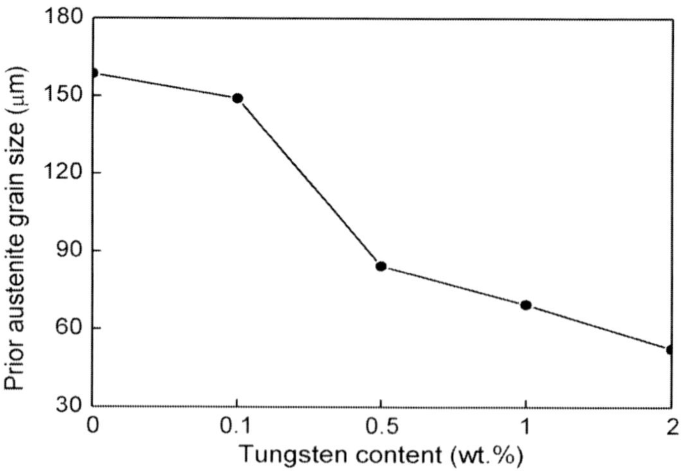

Figure 4.8. Variation of prior austenite grain size with tungsten content. All steels were austenitised at 1230°C for 1 h followed by water quenching [13].

4.3. EFFECT OF TUNGSTEN ON MICROSTRUCTURE

The microstructure of the 0W forging steel is shown in Figure 4.9. It can be seen that the steel without tungsten mainly consists of polygonal ferrite (PF) and pearlite, accompanied by a small quantity of acicular ferrite (AF).

The addition of tungsten has changed the microstructures significantly, as shown in Figure 4.10. Compared with the 0W steel, the amount of AF is increased after 0.1% tungsten addition (Figure 4.10a). When tungsten is increased from 0.1% to 0.5%, allotriomorphic ferrite forms accompanied with an increased amount of AF, as shown in Figure 4.10b.

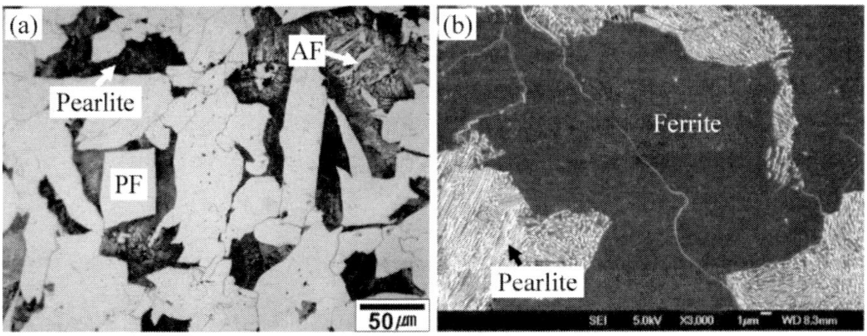

Figure 4.9. Microstructure of the 0W forging steel: (a) OM structure, and (b) SEM image of pearlite.

Allotriomorphic ferrite is the first phase to form on cooling the austenite grains below the A_{e3} temperature. It nucleates at the columnar austenite grain boundaries. Because these boundaries are easy diffusion paths, they become decorated with continuous layers of ferrite. With continuous increasing tungsten from 0.5% to 1%, allotriomorphic ferrite disappears, and the microstructure becomes a mixture of AF and small amount of bainite (Figure 4.10c). When the tungsten content is increased to 2%, the microstructure is still composed of AF and bainite, but is dominated by bainite, as shown in Figure 4.10d. The main difference between AF and bainite is in their nucleation: bainite is nucleated at austenite grain boundaries, while AF is intragranularly nucleated at the nonmetallic inclusions present in the steel [14,15]. The nucleated ferrite in bainite forms groups of parallel plates with the same crystallographic orientation. In contrast, AF leads to a random arrangement of plates, as compared in Figure 4.10b, c and d. The mechanism of tungsten-induced microstructural evolution will be discussed in Section 4.5.

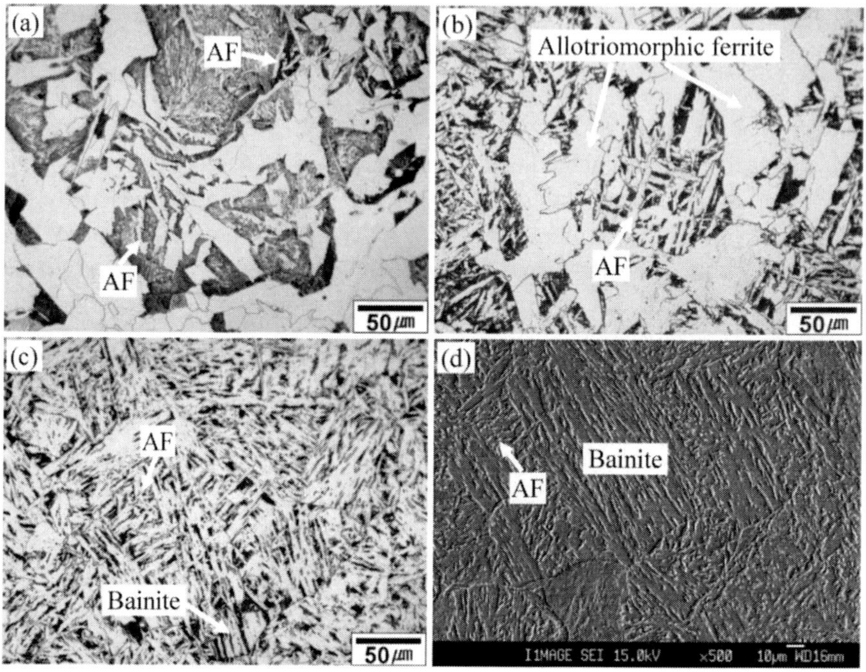

Figure 4.10. OM (a,b,c) and SEM (d) structures of the tungsten-containing forging steels: (a) 0.1W, (b) 0.5W, (c) 1w, and (d) 2W.

4.4. Effect of Tungsten on Interlamellar Spacing of Pearlite

4.4.1. Importance of Interlamellar Spacing of Pearlite

The strength of a ferritic-pearlitic microalloyed steel can be considered to obey one of the following modified relationships [16]:

$$\sigma_y = V_\alpha{}^n \sigma_\alpha + (1 - V_\alpha{}^n)\sigma_p \tag{4.4}$$

$$\sigma_y = V_\alpha \sigma_\alpha + (1 - V_\alpha)\sigma_p + I_{\alpha p} \tag{4.5}$$

$$\sigma_y = V_\alpha \bar{\sigma}_\alpha + (1 - V_\alpha)\bar{\sigma}_p \tag{4.6}$$

where σ_y is the yield strength of ferrite-pearlite aggregate, V_α is the volume fraction of ferrite, n is a constant which incorporates the nonlinear dependence of σ_y on volume fractions of the constituents, σ_α is the yield strength of ferrite, σ_p is the yield strength of pearlite, $I_{\alpha p}$ is the interaction parameter between ferrite and pearlite, $\bar{\sigma}_\alpha$ is the average value of the yield strength of ferrite, and $\bar{\sigma}_p$ is the average value of the yield strength of pearlite.

The strength of pearlite in a steel is considered to be dependent on its interlamellar spacing, and a decrease in the interlamellar spacing of pearlite improves the strength of pearlite, as reported by Modi et al. [17], Gladman et al. [18], and Karlsson and Linden [19]. Therefore, determination of the interlamellar spacing of pearlite in ferritic-pearlitic microalloyed steels becomes essential to evaluate the mechanical properties governed by the constituent phases that are dictated by the solid strengthening by alloying elements, ferrite grain size, and as well as interlamellar spacing of pearlite which depends on the pearlite transformation temperature, steel chemical compositions and other factors.

4.4.2. Measurement of Interlamellar Spacing of Pearlite

The ability and method that can be used to measure the interlamellar spacing of pearlite are vital in any attempt to correlate the spacing with the conditions of pearlite formation. Such correlations will allow for the control and manipulation of the pearlite structure and, therefore, the properties of microalloyed steels. However, measurement of interlamellar spacing is often complicated by the spacing variations within a given pearlite colony, those between different colonies, and those produced by the metallographic sectioning plane with respect to the orientation of the lamellae. A pearlite nodule is composed of multiple colonies, and each colony has parallel lamellae which are orientated differently with respect to lamellae in adjacent colonies. This also exhibits a wide range of interlamellar spacing in different colonies because of the intersection with the polishing plane of pearlite colonies at different angles. Prior to the development of electron microscope, measurements of interlamellar spacing were limited by the resolution capability of the optical microscope or were performed only on lamellae nearly parallel to the plane of polish. With the development of scanning electron microscope and transmission electron microscope, optical microscope methods have been rarely used. A number of measurement procedures have been developed that produce different estimates of the interlamellar spacing of pearlite by many researchers [20-28]. Among these methods, the method recommended by Underwood [23] is the simplest and most universal one to determine the average true interlamellar spacing through measuring the average direct interlamellar spacing. The measurement of the direct interlamellar spacing (S_d) is illustrated in Figure 4.11 [29].

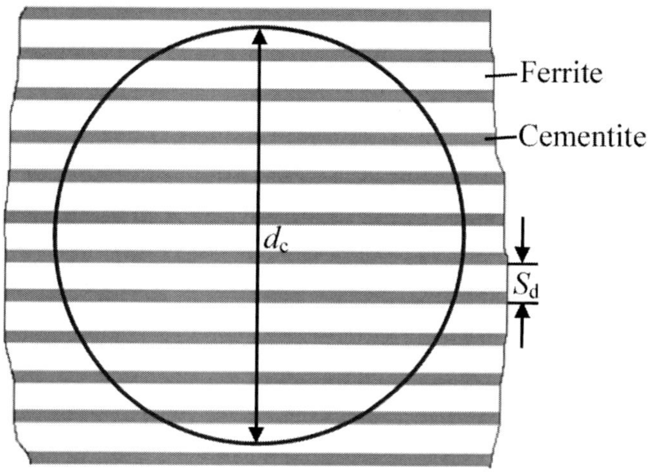

Figure 4.11. Schematic illustration of the measurement of interlamellar spacing of pearlite [29].

For a single pearlite colony, the average direct spacing $\overline{S}_d$ is expressed as:

$$\overline{S}_d = \frac{d_c}{n_c} \tag{4.7}$$

where n_c is the number of cementite lamellae intercepted by the circle diameter d_c. For example, $n_c = 10$ in Figure 4.11.

To determine the interlamellar spacing of pearlite, it is common practice to count the number of cementite interceptions with a straight test line perpendicular to the lamellae. However, because the lamellae intersect the surface at different angles, it is better to determine an average random interlamellar spacing, $\bar{S}_r$, than an average direct interlamellar spacing, $\bar{S}_d$. An average random interlamellar spacing is obtained by determining the number of cementite intercepted by a random circular line, n_r, with the perimeter of πd_c as:

$$\bar{S}_r = \frac{\pi d_c}{n_r} \tag{4.8}$$

In Figure 4.11, $n_r = 20$. For a circular line, n_r generally equals to 2 n_c. Then, Eq. (4.8) can be rewritten as:

$$\bar{S}_r = \frac{\pi}{2}\frac{d_c}{n_c} = \frac{\pi}{2}\bar{S}_d \tag{4.9}$$

Since the average random interlamellar spacing is twice the average true interlamellar spacing, $\bar{S}_t$, the average true interlamellar spacing can, therefore, be easily obtained by:

$$\bar{S}_t = \frac{\bar{S}_r}{2} = \frac{\pi}{4}\bar{S}_d \tag{4.10}$$

In practice in ferritic-pearlitic microalloyed steels, the magnification of an electron micrograph should be high enough so that most of the pearlite grains are larger than the apparent test grid area. The test grid should not overlap the pearlite and cover ferrite phase. However, as the magnification is raised and the number of intersections, n_r, per field decreases, measurement accuracy will decrease.

In order to obtain the best accuracy, the lowest feasible magnification, which must be high enough for adequate resolution and accurate counting but is not higher than necessary so that n_r may be maximised, should be used. Micrographs that can present clearly the parallel lamellae structure of pearlite should be used in the measurement of interlamellar spacing, as shown in

Figure 4.12a. An image shown in Figure 4.12b that exhibits only the cross section of the parallel lamellae is generally not suitable to perform an accurate measurement of the interlamellar spacing of pearlite by using the aforementioned method.

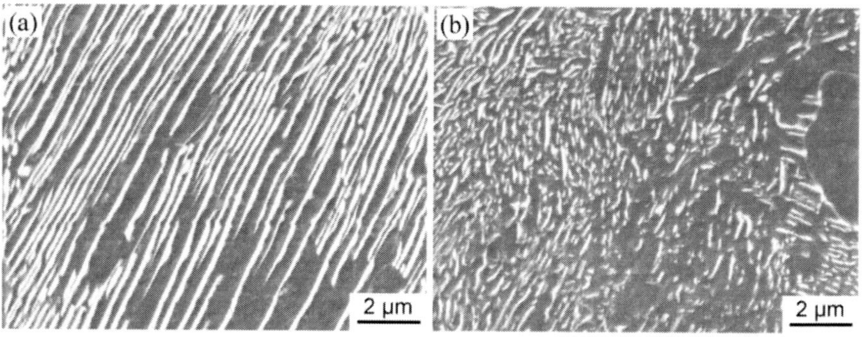

Figure 4.12. SEM of pearlite of the 0W steel: (a) parallel structure of lamellae, and (b) cross section of parallel lamellae.

Measurements should be made without bias. One should not deliberately attempt to select fields that appear to be typical, minimum or maximum in the interlamellar spacing of pearlite. Measurement accuracy increases as the number of measurements increases. So the measurements should be repeated on as many fields as possible to obtain a good estimate of the interlamellar spacing of pearlite.

4.4.3. Variation of Interlamellar Spacing of Pearlite after Tungsten Addition

Figure 4.13 shows the typical morphologies of pearlite observed in the tungsten-free and tungsten-containing microalloyed steels. As indicated in the SEM images, the interlamellar spacing of pearlite has found to be decreased after tungsten addition. By performing measurements on 50 random SEM images at magnifications from ×5000 to ×7000 without bias using the method introduced in Section 4.4.2, the average direct interlamellar spacing $\overline{S}_{\mathrm{d}}$ and average true interlamellar spacing $\overline{S}_{\mathrm{t}}$ of pearlite of the tungsten-free steel are obtained to be 238 and 187 nm, respectively. Using the same method, the measured $\overline{S}_d$ and $\overline{S}_t$ values of the tungsten-containing steels are: $\overline{S}_d = 215$ nm and $\overline{S}_t = 169$ nm for the 0.1W steel; and $\overline{S}_d = 201$ nm and $\overline{S}_t = 158$ nm for the 0.5W steel, as presented in Table 4.2 [29]. Therefore, the addition of tungsten in microalloyed steels has a positive effect on the decrement of interlamellar spacing of pearlite.

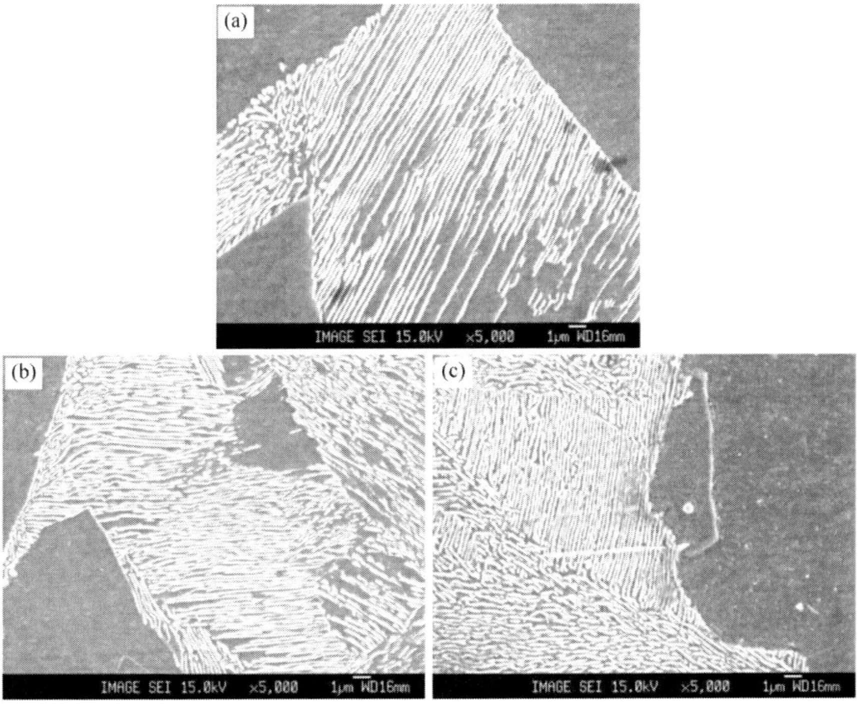

Figure 4.13. SEM images of pearlite of the (a) 0W, (b) 0.1W and (c) 0.5W steels.

Table 4.2. Interlamellar spacing of pearlite of the 0W, 0.1W and 0.5W steels [29]

Steel	$\overline{S}_d$ (nm)	$\overline{S}_t$ (nm)
0W	238	187
0.1W	215	169
0.5W	201	158

4.5. Mechanism of Tungsten-Induced Microstructural Evolution

The term "acicular ferrite (AF)" was first used by Grong [30] and Abson [31] in 1986. AF usually forms in forged and welded steels, and the term "acicular" means shaped and pointed like a needle, but the true shape is that of a lenticular plate. AF nucleation takes place inside the austenite grains on nonmetallic inclusions. Nonmetallic inclusions can be oxides or other compounds, such as carbides, nitrides and carbonitrides in microalloyed steels, but the important point is that they may stimulate AF [32]. Nonmetallic inclusions are necessary for the formation of AF. Steel which contains relatively large numbers of inclusions can provide abundant sites for AF nucleation, and then higher content of AF in the microstructure can be obtained. The nucleation of a single plate on an inclusion can in turn stimulate others to nucleate autocatalytically, so that a one-to-one correspondence between the number of active inclusions and the number of AF plates is not expected [33]. The plates of AF nucleate heterogeneously on small nonmetallic inclusions and radiate in many different directions from these point nucleation sites. AF does not grow in sheaves because their development is stifled by the impingement of plates nucleated independently at adjacent sites [34]. Since AF grows by a displacive mechanism, plates of AF are confined to austenite grains, where they grow because the coordinated movement of atoms associated with the displacive transformation mechanism cannot be sustained across grain boundaries [35]. A schematic diagram of AF formation is presented in Figure 4.14 [13].

The displacive mechanism of AF nucleation relies on the existence of arrays of dislocations. These arrays are generated in the proximity of nonmetallic inclusions due to stresses caused by differential thermal expansion. Such stresses are more difficult to accommodate for larger inclusions, making large inclusions more potent nucleation sites. Ricks et al. [33] studied the mechanism of AF nucleation based on classical heterophase fluctuation theory, and he found that the activation energy (ΔG^*) for a fluctuation which is large enough to stimulate a critical nucleus depends on the inverse square of the chemical driving force, $\Delta G^* \propto \Delta G_{chem}^{-2}$. According to this theory, it is clear that larger size nonmetallic inclusions are more effective for heterogeneous AF nucleation. Beside nonmetallic inclusions, another factor, the austenite grain size, is also responsible for the formation of AF. AF is intragranularly nucleated bainite, and it is possible to switch between these two morphologies by controlling the nucleation sites. The appearance of AF is only different from that of bainite because it nucleates intragranularly in steels containing a greater number density of inclusions than austenite grain surface nucleation sites [34]. Bainite forms when the austenite grain size is small because nucleation then predominates at the grain boundaries. Subsequent growth then swamps the interiors of austenite grains, preventing the development of AF. A microstructure with large austenite grains has a better chance of transformation to AF at the expense of bainite because the number density of inclusions becomes large enough relative to the boundary nucleation sites.

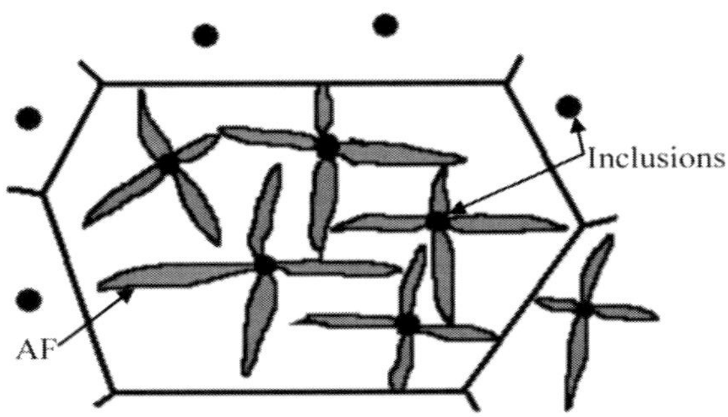

Figure 4.14. Schematic diagram of AF formation [13].

According to the analysis results on precipitates in Chapter 3, and prior austenite grains in Section 4.2, no noticeable increase in the amount of precipitates after tungsten addition compared to the specimen without tungsten (Figures 3.8 and 3.9), and the average sizes of both austenite grains and precipitates decrease with an increase of tungsten content (Figures 3.10 and 4.8). Obviously, these factors are not beneficial to the formation of AF. Another factor, allotriomorphic ferrite, is then considered to analyse the effects of tungsten on the development of AF in mixed microstructures.

Allotriomorphic ferrite is the first phase to form on cooling the austenite grains below the A_{e3} temperature [36]. Nucleation takes place at the columnar austenite grain boundaries. Because these boundaries are easy diffusion paths, they become decorated with thin, continuous layers of ferrite, as illustrated in Figure 4.15 [13]. As shown in Figure 4.9, allotriomorphic ferrite can scarcely be found in the 0W steel. Conversely, the amount of allotriomorphic ferrite increases gradually with increasing tungsten from 0 to 0.5%, and the addition of 0.5% tungsten induces the highest volume fraction of allotriomorphic ferrite to be formed in the microstructure (Figure 4.10). Figure 4.16 shows an SEM image of allotriomorphic ferrite formed in the 0.5W steel at a higher magnification [13]. It can be seen that the shape of the allotriomorphic ferrite is very intact, which means the austenite grain surfaces are probably entirely covered by allotriomorphic ferrite. When the tungsten content is increased to 1% or greater, however, no allotriomorphic ferrite can be found in the microstructure, as shown in Figure 4.10c and d. Therefore, when the tungsten content is no greater than a certain value, it can favour the formation of allotriomorphic ferrite.

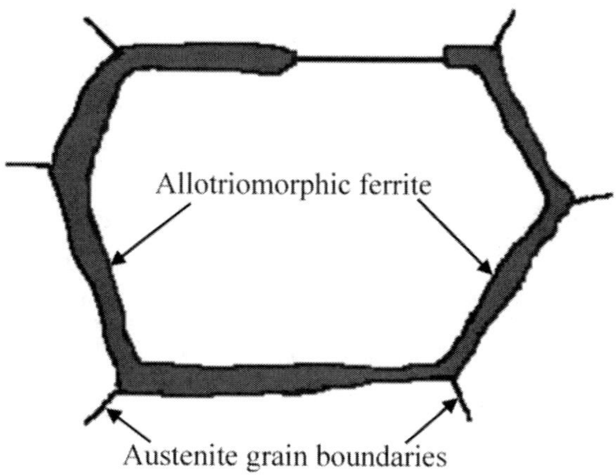

Figure 4.15. Allotriomorphic ferrite formed at austenite grain boundaries [13].

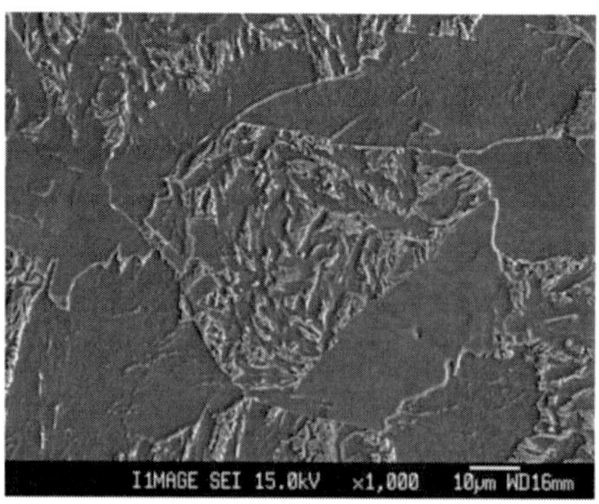

Figure 4.16. SEM image of allotriomorphic ferrite formed in the 0.5W steel [13].

AF is intragranularly nucleated bainite, and the formation of allotriomorphic ferrite is favourable to the development of AF instead of bainite because the austenite grain surface sites for bainite nucleation are covered by allotriomorphic ferrite, and the austenite grain boundaries are then not free to nucleate bainite [37]. As indicated in Figures 4.9 and 4.10, the amount of allotriomorphic ferrite increases gradually with tungsten when its content is less than 0.5%, the volume fraction of AF consequently increases. When the content of tungsten is high (> 0.5%), the formation of bainite in the microstructure may be related to the reduction in the coverage of austenite grain boundaries by layers of allotriomorphic ferrite, so that the austenite grain boundaries are then freed for bainite nucleation. When containing tungsten less than 0.5%, the steel hardenability is low enough to ensure that the austenite grain surfaces are completely covered by uniform layers of allotriomorphic ferrite, thereby rendering them useless for bainite nucleation, but consequently allowing the development of AF by intragranular transformation. As the tungsten content increases, some of the austenite grain surfaces are left bare and become available for the formation of bainite sheaves, as soon as the temperature falls within the bainite transformation range. Also, at relatively high tungsten content, the sizes of both austenite grains and precipitates are small, favouring the formation of bainite instead of AF. The mechanism of the transition from AF to bainite with an increase of the tungsten content is schematically illustrated in Figure 4.17 [13].

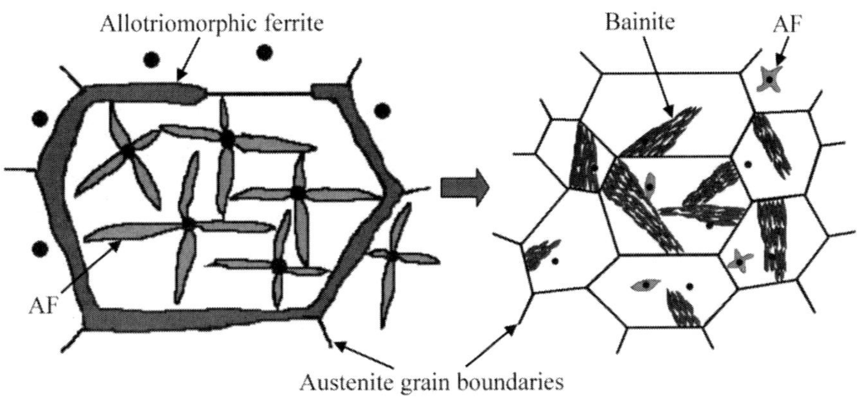

Figure 4.17. Schematic illustration of the mechanism of the transition from AF to bainite with increasing the tungsten content [13].

4.6. Summary

In this chapter, the dependences of austenite grain size of a tungsten-free microalloyed steel on austenitising temperature and holding time are analysed. The effects of tungsten addition on the austenite grain size, microstructural constituents and feature, and interlamellar spacing of pearlite of several microalloyed forging steels are systematically described. The mechanism of tungsten-induced microstructural evolution of microalloyed steels is discussed.

Austenitising temperature affects significantly the austenite grain size. Dissolution of vanadium-containing precipitates plays a key role in the abnormal growth of austenite grains when the temperature is increased from 1100 to 1150°C. Further growth in austenite grains at 1200°C should be attributed to the dissolution of titanium-containing particles and the reduction of the amount of precipitates. Holding time is found to have little influence on the austenite grain size.

The addition of tungsten in microalloyed steels induces refined austenite grain size. When the tungsten content is equal to or less than 0.5%, the addition of tungsten favours the formation of allotriomorphic ferrite, which subsequently promotes the development of AF in microalloyed forging steels. Tungsten alloying in microalloyed steels has a positive effect on the decrement of interlamellar spacing of pearlite.

References

[1] G. Krauss, *Steels: Processing, Structure, and Performance*, ASM International, 2005.

[2] Q.Y. Sha, G.J. Huang, J. Guan, X.J. Ma, D.H. Li, A new route for identification of precipitates on austenite grain boundary in an Nb-V-Ti microalloyed steel, *Journal of Iron and Steel Research, International*, 18(8) (2011) 53-57.

[3] L.Y. Lan, C.L. Qiu, D.W. Zhao, X.H. Gao, L.X. Du, Dynamic and static recrystallization behavior of low carbon high niobium microalloyed steel, *Journal of Iron and Steel Research, International*, 18(1) (2011) 55-60.

[4] K. Banerjee, M. Militzer, M. Perez, X. Wang, Nonisothermal austenite grain growth kinetics in a microalloyed X80 linepipe steel, *Metallurgical and Materials Transactions A*, 41A (2010) 3161-3172.

[5] J.G. Jung, J.S. Park, J. Kim, Y.K. Lee, Carbide precipitation kinetics in austenite of a Nb-Ti-V microalloyed steel, *Materials Science and Engineering A*, 528 (2011) 5529-5535.

[6] A.G. Kostryzhev, A.A. Shahrani, C. Zhu, S.P. Ringer, E.V. Pereloma, Effect of deformation temperature on niobium clustering, precipitation and austenite recrystallization in a Nb-Ti microalloyed steel, *Materials Science and Engineering A*, 581 (2013) 16-25.

[7] L. Zhang, Z.X. Yuan, S.H. Song, T.H. Xi, Q. Wang, Austenite grain growth in heat affected zone of Zr-Ti bearing microalloyed steel, *Journal of Iron and Steel Research, International*, 19(2) (2012) 73-78.

[8] M. Tarafder, P. Sinha, A. Kundu, M. Strangwood, C. Davis, Fractal based correlations for Nb microalloyed steel undergoing static recrystallization, *Materials Characterization,* 85 (2013) 92-99.

[9] M. Esmailian, The effect of cooling rate and austenite grain size on the austenite to ferrite transformation temperature and different ferrite morphologies in microalloyed steels, *Iranian Journal of Materials Science and Engineering*, 7(1) (2010) 7-14.

[10] J. Moon, J. Lee, C. Lee, Prediction for austenite grain size in the presence of growing particles in the weld HAZ of Ti-microalloyed steel, *Materials Science and Engineering A*, 459 (2007) 40-46.

[11] J. Zhao, Z. Jiang, C.S. Lee, Enhancing impact fracture toughness and tensile properties of a microalloyed cast steel by hot forging and post-forging heat treatment processes, *Materials and Design*, 47 (2013) 227-233.

[12] J. Zhao, J.H. Lee, Y.W. Kim, Z. Jiang, C.S. Lee, Enhancing mechanical properties of a low-carbon microalloyed cast steel by controlled heat treatment, *Materials Science and Engineering A*, 559 (2013) 427-435.

[13] J. Zhao, T. Lee, J.H. Lee, Z. Jiang, C.S. Lee, Effects of tungsten addition on the microstructure and mechanical properties of microalloyed forging steels, *Metallurgical and Materials Transactions A,* 44A (2013) 3511-3523.

[14] J.R. Yang, H.K.D.H. Bhadeshia, The orientation relationships between adjacent plates of acicular ferrite in steel weld deposits, *Materials Science and Technology*, 5 (1989) 93-97.

[15] I. Madariaga, I. Gutierrez, H.K.D.H. Bhadeshia, Acicular ferrite morphologies in a medium-carbon microalloyed steel, *Metallurgical and Materials Transactions A*, 32 (2001) 2187-2197.

[16] K.K. Ray, D. Mondal, The effect of interlamellar spacing on strength of pearlite in annealed eutectoid and hypoeutectoid plain carbon steel, *Acta Metallurgica et Materialia*, 39 (1991) 2201-2208.

[17] O.P. Modi, N. Deshmukh, D.P. Mondal, A.K. Jha, A.H. Yegneswaran, H.K. Khaira, Effect of interlamellar spacing on the mechanical properties of 0.65%C steel, *Materials Characterization*, 46 (2001) 347-352.

[18] T. Gladman, I.D. Mcivor, F.B. Pickering, Some aspects of the structure-property relationships in high-carbon ferrite-perlite steels, *Journal of the Iron and Steel Institute*, 210 (1972) 916-930.

[19] B. Karlsson, G. Lindén, Plastic deformation of eutectoid steel with different cementite morphologies, *Materials Science and Engineering*, 17 (1975) 153-164.

[20] G.E. Pellissier, M.F. Hawekes, W.A. Johnson, R.F. Mehl, The interlamellar spacing of pearlite, *Transactions ASM*, 30 (1942) 1049-1086.

[21] J.W. Cahn, R.L. Fullman, On the use of lineal analysis for obtaining particle size distribution functions in opaque samples, *Transactions AIME*, 206 (1956) 610-612.

[22] G. Birkbeck, T.C. Wells, Pearlite morphology in three low-carbon steels, *Transactions AIME*, 242 (1968) 2217-2220.

[23] E.E. Underwood, *Quantitative Stereology*, Addison-Wesley Publishing Co., Reading, Massachusetts, 1970.

[24] A. Roósz, Z. Gácsi, M.K. Baan, A simple method for determining the true interlamellar spacing, *Metallography*, 13 (1980) 299-306.

[25] G.F. Vander Voort, Measurement of the interlamellar spacing of pearlite, *Metallography*, 17 (1984) 1-17.

[26] J.F. Tilbury, T.D. Motishaw, G.D.W. Smith, Measurements of the spread in pearlite spacings, *Metallography*, 19 (1986) 243-246.

[27] H.S. Fong, Determining true pearlite lamellar spacings from observed apparent spacings, *Metallography*, 23 (1989) 173-188.

[28] F.G. Caballero, C. García de Andrés, C. Capdevila, Characterization and morphological analysis of pearlite in a eutectoid steel, *Materials Characterization*, 45 (2000) 111-116.

[29] J. Zhao, Y.W. Kim, J.H. Lee, J.M. Lee, H.S. Chang, C.S. Lee, Effect of tungsten addition on the mechanical properties and corrosion resistance of S355NL forging steel, *Metals and Materials International*, 18 (2012) 217-223.

[30] Ø. Grong, D.K. Matlock, Microstructural development in mild and low-alloy steel weld metals, *International Metals Reviews*, 31 (1986) 27-48.

[31] D.J. Abson, R.J. Pargeter, Factors influencing as-deposited strength, microstructure, and toughness of manual metal arc welds suitable for C-Mn steel fabrications, *International Metals Reviews*, 31 (1986) 141-194.

[32] Y. Ito, M. Nakanishi, Study on Charpy impact properties of weld metal with submerged arc welding, *Sumitomo Search*, 15 (1976) 42-62.

[33] R.A. Ricks, P.R. Howell, G.S. Barritte, The nature of acicular ferrite in HSLA steel weld metals, *Journal of Materials Science*, 17 (1982) 732-740.

[34] J.R. Yang, H.K.D.H. Bhadeshia, Thermodynamics of the acicular ferrite transformation in alloy-steel weld deposits, in: S.A. David (Ed.), *Advances in Welding Technology and Science*, ASM, Metals Park, Ohio, 1987.

[35] H.K.D.H. Bhadeshia, *Bainite in Steels*, 2nd ed., The University Press, Cambridge, London, 2001.

[36] H.K.D.H. Bhadeshia, R.W.K. Honeycombe, *Steels: Microstructure and Properties*, 3rd ed., Elsevier Butterworth-Heinemann, Oxford, 2006.

[37] S.S. Babu, H.K.D.H. Bhadeshia, The transition from bainite to acicular ferrite in reheated Fe-Cr-C weld deposits, *Materials Science and Technology*, 6 (1990) 1005-1020.

Chapter 5

EFFECT OF TUNGSTEN ON MECHANICAL PROPERTIES

ABSTRACT

In Chapter 4, the addition of tungsten in microalloyed steels is found to have a significant effect on the microstructure. This chapter describes the mechanical properties of microalloyed steels with different tungsten contents, and analyses how the microstructural evolution affects the mechanical properties after addition of tungsten. The importance of mechanical properties in the evaluation of microalloyed steels is at first stated. Mechanical tests including tensile, impact and hardness tests are introduced to help understanding the testing procedures and the assessment criteria on the mechanical properties. The effects of tungsten on the hardness, room/low-temperature impact toughness, and room/low-temperature tensile properties are analysed. The dependence of tensile strength on the testing temperature is quantified. At room temperature, dislocations in the matrix of microalloyed steels can readily move and multiply to produce a large strain. However, below room temperature, dislocation movement is inhibited by the Peierls force, and an early fracture may occur. These aspects of deformation behaviour and the mechanism of tungsten-induced variation of room/low-temperature mechanical properties are discussed in following sections.

5.1. GENERAL CONSIDERATION

The quality that is required from any material depends on its use. Microalloyed steels are often subjected to external forces (loads) when they are used. Basically, external forces can be divided into three categories, i.e., tension, compression and shear. When an external force tends to cause a steel to change shape, the steel is said to be stressed. If the force applied is so strong that the shape is actually deformed permanently, the steel is regarded to be strained or distorted. Mechanical properties of microalloyed steels are defined as the reaction of the steels to certain types of external forces. The mechanical properties of microalloyed steels determine the range of usefulness of the steels and establish the expected services, and are of vital importance in the determination of their fabrication and practical applications. Mechanical properties are also employed to help specify and identify microalloyed steels with specific purposes.

The most common mechanical properties considered in the evaluation of the performance of microalloyed steels are strength, ductility, fracture toughness and hardness. In some cases, creep resistance and fatigue limits of microalloyed steels are also considered based on their application environments. Mechanical properties measured of microalloyed steels are generally associated with the way that the steels behave when they are subjected to various states of external forces. Testing methods are important to determine mechanical properties and to ensure that standards are adhered to. Methods of test determine the yield strength (YS), ultimate tensile strength (UTS) and ductility through tensile test, fracture toughness through impact test, and hardness through resistance to penetration of the surface by a hard object. Test must be accurate to determine the mechanical properties of a microalloyed steel and to ensure adherence to standards such as ASTM.

5.2. Mechanical Test of Microalloyed Steels

5.2.1. Tensile Test

Tensile test, also known as tension test, is a fundamental materials science test in which a sample is subjected to a controlled tension until failure. The results obtained from a tensile test are commonly used to evaluate a steel for a specific application, for quality control, and to predict how the steel will react under other types of forces being applied in practical situations such as in bridges or vehicle structures. A lot of information can be learned about a steel from a tensile test. As the steel is stretched until failure occurs, a comprehensive tensile profile will result producing a curve showing how it reacted to the forces being applied. This curve is usually an engineering stress-strain diagram. The values of both engineering stress and engineering strain are generally calculated automatically by collecting the force applied and the elongation of the specially shaped specimen suffered during a tensile test through a computer control system. Properties that can be directly obtained from an engineering stress-strain curve are Young's modulus, YS, UTS, fracture strength and maximum strain. Other tensile properties including strain hardening behaviour and reduction in area can be acquired via relevant analysis on the tensile curve and observation on the specimen before and after fracture.

Uniaxial tensile test is the most commonly used testing method for evaluating the mechanical properties of microalloyed steels and other steel grades that are isotropic. A variety of tensile machines are available for tensile tests, among which universal tensile testing machines are popular and have been applied in many scientific research institutes, universities and large manufacturing companies such as Baosteel, China and POSCO, South Korea, for obtaining the mechanical characteristics of steels and other metal alloys. One type of universal tensile testing machine is illustrated in Figure 5.1.

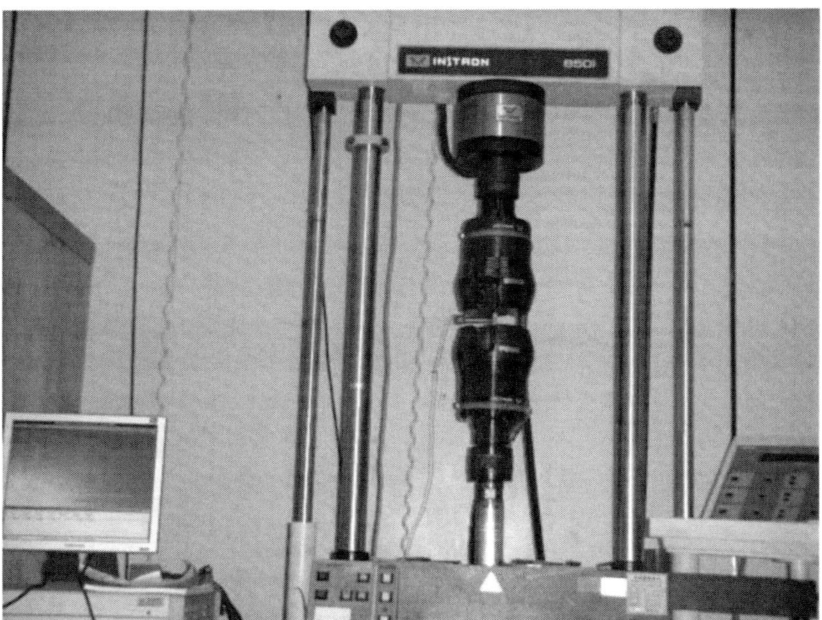

Figure 5.1. INSTRON 8501 universal tensile testing machine.

Stress and Strain

To conduct a tensile test, a steel specimen is usually machined to be the form of a round bar or a flat plate following relevant standard such as ASTM E8 [1], as illustrated in Figure 5.2. In the figure, L_t is the total length, L_g is the length of grip section, L_0 is the gage length, L_r is the length of reduced section, R is the radius fillet, d_0 is the diameter of the reduced section of the round bar specimen, t_0 is the thickness of the flat plate specimen, and w_0 is the width of the reduced section of the flat plate specimen.

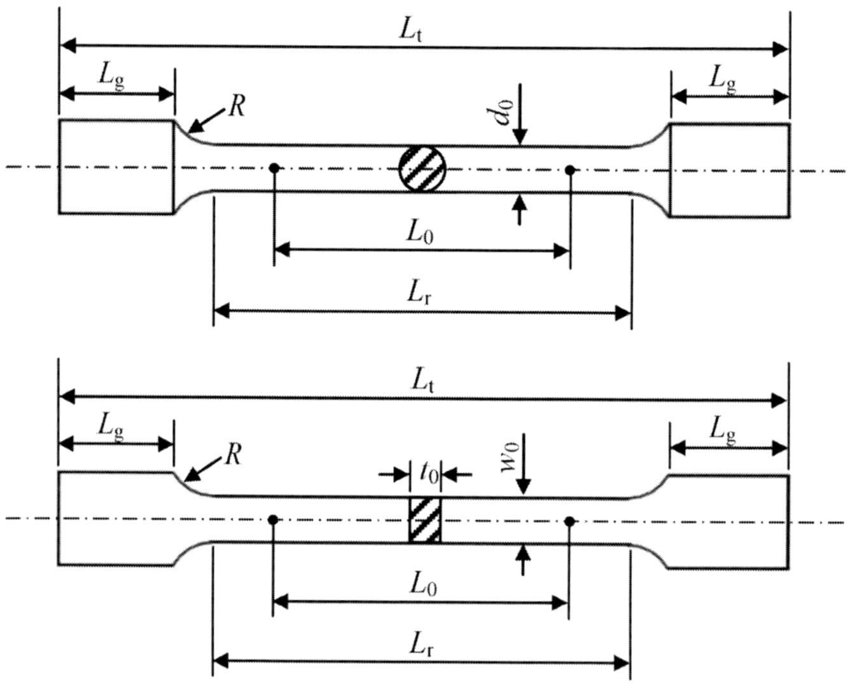

Figure 5.2. Typical tensile specimens with the shapes of round bar and flat plate.

Stress-strain curve is a representation of the performance of a specimen as the applied load is increased until fracture occurs. Two kinds of stress-strain curves, i.e., "engineering stress-strain curve" and "true stress-strain curve" are usually employed to evaluate the tensile properties of a tested specimen.

The engineering stress, σ_e, is obtained by dividing the load applied, P, by the original area of the cross section of the specimen, A_0:

$$\sigma_e = \frac{P}{A_0} \tag{5.1}$$

The engineering strain, ε_e, is the average linear strain which is obtained by dividing the elongation of the gage length of the specimen, ΔL, by its gage length, L_0:

$$\varepsilon_e = \frac{\Delta L}{L_0} = \frac{L - L_0}{L_0} = \frac{L}{L_0} - 1 \tag{5.2}$$

where L is the gage length of the specimen after tensile deformation.

Engineering stress and strain are independent of the dimensions of the specimen tested. The engineering stress-strain curve has the same shape as the load-elongation curve because both the engineering stress and engineering strain are obtained by dividing the load and elongation by constant factors A_0 and L_0, respectively. The two kinds of curves are therefore usually used interchangeably.

In the engineering stress-strain curve, there is a stress drop beyond the UTS, indicating no increase in load is essential for continued deformation. This is not the case, however, as indicated from the true nature of a tensile test. The UTS in the engineering stress-strain curve results from tensile instability, not from a decrease in the strength of the tested material. The drop of stress in the engineering stress-strain curve is caused due to the calculations based on the original cross sectional area of the tested specimen, but actually necking has been initiated beyond the strain corresponds to the UTS. In contrast of the engineering stress which is defined with respect to the original area of cross section of the specimen, the true stress, σ_t, is calculated based on the instantaneous area, i.e., the real area, at any instant of load, A_i. The true stress is defined below:

$$\sigma_t = \frac{P}{A_i} = \frac{P}{A_0} \cdot \frac{L}{L_0} = \sigma_e \cdot (\varepsilon_e + 1) \tag{5.3}$$

If dL is the amount by which the length L changes, the true strain can then be defined as the sum of all the instantaneous engineering strains as:

$$\varepsilon_t = \int_{L_0}^{L} \frac{dL}{L} = \ln \frac{L}{L_0} = \ln(\varepsilon_e + 1) \tag{5.4}$$

In steels, the tensile instability generally corresponds to the onset of necking in the tested specimen. Necking results from strain localisation, and thus true strain cannot be calculated directly from elongation once necking is initiated. Therefore, if a tensile test of a steel is used to understand and predict the metallurgical response during the deformation associated with fabrication processes, true stress-strain curve is preferred. The true stress and true strain are practically indistinguishable from the engineering stress and engineering strain at small deformations. As the strain becomes large and the cross sectional area of the specimen decreases, the true stress can be much larger than the engineering stress. A comparison of engineering and true stress-strain curves is summarised graphically in Figure 5.3.

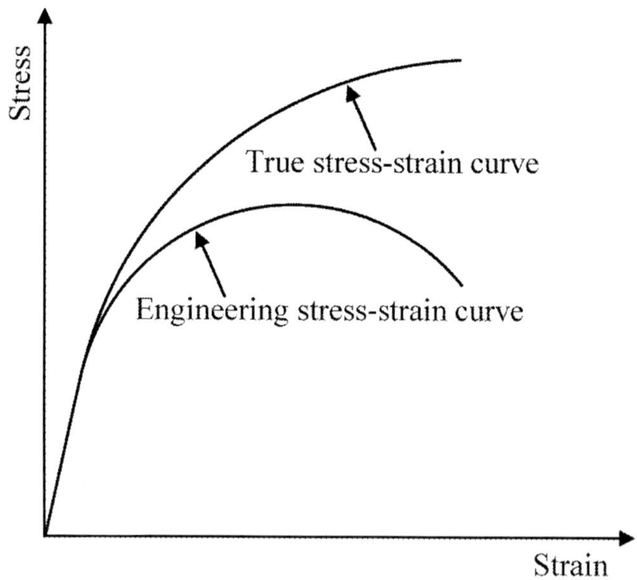

Figure 5.3. Comparison of engineering and true stress-strain curves.

In the International System of Units (SI), the unit of stress is Newton per square meter, which is also called Pascal. An example is given below to help understand the calculation process of both engineering and true stresses and strains from a tensile test.

Consider a round bar tensile specimen with an original diameter of cross section of $d_0 = 6\ \text{mm}$ and gage length of $L_0 = 30\ \text{mm}$ reaches a load of $P = 15 \times 10^3\ \text{N}$ at which the gage length of the specimen becomes $L = 36\ \text{mm}$, the engineering stress can be calculated based on Eq. (5.1) as:

$$\sigma_e = \frac{P}{A_0} = \frac{15 \times 10^3}{\pi (6 \times 10^{-3})^2 / 4} = 530.5 \times 10^6\ \text{Pa} = 530.5\ \text{MPa}$$

The engineering strain can be obtained according to Eq. (5.2) as:

$$\varepsilon_e = \frac{L}{L_0} - 1 = \frac{36}{30} - 1 = 0.2$$

Based on Eqs. (5.3) and (5.4), the true stress and true strain can be calculated respectively as follows:

$$\sigma_t = \sigma_e \cdot (\varepsilon_e + 1) = 530.5 \times (0.2 + 1) = 636.6\ \text{MPa}$$

$$\varepsilon_t = \ln(\varepsilon_e + 1) = \ln(0.2 + 1) = 0.18$$

Young's Modulus

Young's modulus, also known as modulus of elasticity, is a measure of the stiffness or rigidity of a material when a force is applied to it. It can be obtained from the slope of its engineering stress-strain curve in the elastic deformation region in which Hooke's Law is obeyed [2]:

$$E = \frac{\sigma_e}{\varepsilon_e} \tag{5.5}$$

where E is the Young's modulus. Figure 5.4 illustrates how the value of E is determined. The Young's modulus applies only in the linear region of the stress-strain curve. If a steel specimen is loaded within such linear region, the steel will return to its original state when the load is released, i.e., the steel is deformed elastically.

Strength

The parameters that are used to describe the strength of a steel from the engineering stress-strain curve include YS, UTS and fracture strength, as schematically illustrated in Figure 5.5.

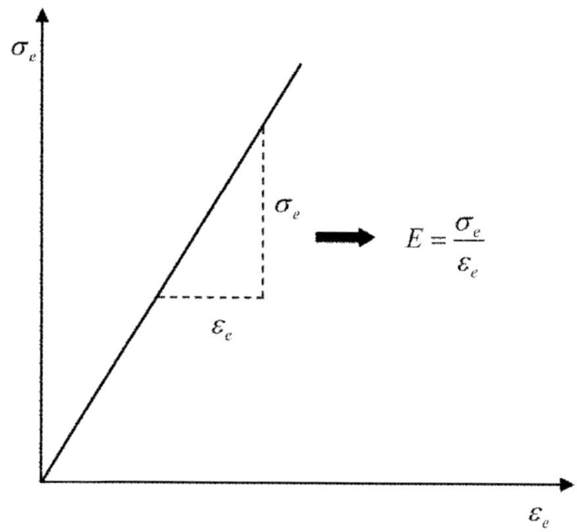

Figure 5.4. Illustration of the determination of Young's modulus.

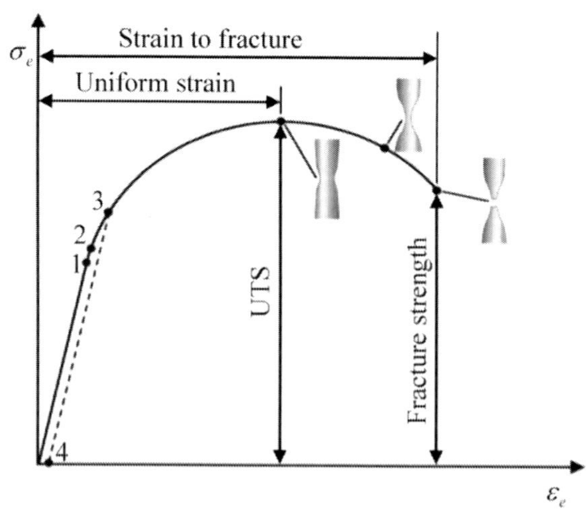

Figure 5.5. Engineering stress-strain curve showing critical stress and strain: "1"-proportional limit, "2"-elastic limit, "3"-offset YS, and "4"-offset strain.

There is a gradual transition from elastic to plastic deformation behaviour with the increase of stress, and the transition point at which plastic deformation starts is difficult to detect with precision. Three criteria are generally used to determine the initiation of yielding: proportional limit, elastic limit and the YS.

Proportional limit. Up to this amount of stress, the stress depends linearly on the strain, i.e., obeys Hooke's Law, and the slope of the stress-strain curve equals to the Young's modulus, as shown at point "1" in Figure 5.5.

Elastic limit. This limit indicates the greatest stress that a steel can withstand without any measurable permanent strain remaining when the load is completely released. Beyond this limit, Hooke's Law no longer applies and some permanent deformation occurs, as shown at point "2" in Figure 5.5.

YS. This strength is the stress required to produce a small specified amount of plastic deformation. Beyond this strength the stress-strain curve levels off and plastic deformation starts to occur, as shown at point "3" in Figure 5.5. In some steels, continuous yielding takes place during the tensile test. For these steels, the yield strength is generally defined as the offset determined by the stress corresponding to the intersection of the stress-train curve offset by a specified strain. The offset is commonly set at 0.2% or 0.1% of the strain, as shown at point "4" in Figure 5.5.

UTS is the maximum engineering stress that a steel can withstand while being stretched or pulled before necking phenomenon takes place during a tensile test. UTS is the value that most frequently quoted from the results of a tensile test, and the highest point of the engineering stress-strain curve indicates the value of UTS, as illustrated in Figure 5.5.

UTS is an intensive property, and its value does not depend on the size of the tested specimen. Other factors such as the preparation of specimen, the presence or otherwise of surface defects, and the testing temperature, however, may affect the value of UTS. UTS is of little fundamental significance with regard to the strength of a steel, and its value should be regarded as a measure of the maximum load that the steel can withstand under very restrictive conditions of uniaxial loading. However, this value bears little relationship to the useful strength of the steel under more complex conditions of stress that are usually encountered.

Fracture strength is the stress at which a specimen fails via fracture. In a tensile test, the corresponding stress of the final recorded point in the engineering stress-strain curve is defined as the fracture strength, as shown in Figure 5.5.

Ductility

The ductility of a steel is a measure of the degree to which the steel can be plastically deformed without fracture. Ductility is an important factor should be considered in steel forming operations such as forging, rolling, drawing and extrusion. A steel with high ductility will be able to be stretched into long, thin wires without fracture. Contrarily, a steel with low ductility may only experience very little or no plastic deformation upon fracture. Ductility provides an indication of how visible overload damage to a steel product might become before the product fractures.

The ductility of a steel is generally established by measuring either the elongation to fracture or the reduction of area at fracture in a tensile test. Elongation is a measure of the percentage change in axial length divided by the original length of the specimen or portion of the specimen. The measured elongation from a tensile specimen depends on the gage length over which the measurement is taken. Increasing the gage length of a specimen will decrease the elongation to fracture. This is because after necking forms all subsequent deformation will be concentrated in the vicinity of the necked region. The behaviour around the neck is the same regardless of the length of the tested specimen. The shorter the gage length, the greater the localised deformation at the neck will factor into the calculation. Thus a relatively large fraction of the specimen's length is deforming during the tensile test if the gage length is short. In a longer specimen, the portion away from the neck does not continue to deform after the onset of necking, and a smaller elongation of the specimen that contributes to the total deformation will be obtained. Therefore, to report the value of elongation, the gage length of the specimen applied in the tensile test should be provided. Be sure to use the same gage length when comparing elongation among different steel grades.

Reduction of area is the change in cross sectional area divided by the original cross sectional area. This change is measured in the necked region of the tested specimen. Like elongation, it is also expressed as a percentage. Reduction of area is an important parameter in steel forming operations especially in the deep drawing fabrication process of the production of cups and other items.

5.2.2. Impact Test

Impact test is a testing method to determine the ability of a steel to resist high-rate loading. Impact toughness obtained from an impact test is a measure of how much energy a steel can absorb without rupturing. It is a very important mechanical property for steel structures and products subject to impact loading and especially in shipping industry, offshore industry, aviation industry, pressure vessels and bridges in cold and stormy conditions.

There are primarily two types of impact tests, i.e., Charpy impact test [3] and Izod impact test [4], both of which are named respectively after their inventors. Both tests involve fracturing a notched specimen with standard dimension by a single overload event due to the impact of the pendulum. The impact toughness is determined by measuring the energy absorbed in the fracture of the specimen. This is simply obtained by recording the height at which the pendulum is released and the height to which the pendulum swings after it has struck the specimen. The height of the pendulum times the weight of the pendulum produces the potential energy and the difference in potential energy of the pendulum at the start and the end of the test is equal to the absorbed energy. In practice, the absorbed energy is generally calculated automatically by collecting the relevant information during impact test through a computer control system.

The toughness of a steel is greatly affected by the servicing temperature. By applying impact tests at different temperatures, a graph of absorbed impact energy as a function of temperature for a given steel can be plotted, from which the ductile-to-brittle transition (DBT) can be clearly observed. The DBT temperature is the boundary between brittle and ductile fracture behaviours, and this temperature is often an extremely important consideration in the section of a steel for specified applications.

Practically, the determination of the DBT temperature is somewhat difficult because the transition is not sharp. A relatively exact DBT temperature may be empirically defined in several ways, e.g., a specified absorbed energy, change in aspect of fracture (such as the fracture surface containing 50% of cleavage appearance), etc.

5.2.3. Hardness Test

Hardness is a measure of how resistant a steel is to various kinds of permanent shape change when a force is applied. Hardness is not a fundamental property. Its value is arbitrary, and there is no absolute standard of hardness. Hardness has no quantitative value except in terms of a given load applied in a specified manner for a specified duration and a specified penetrator shape.

There are several testing methods for measuring the hardness of a steel including static indentation hardness test, dynamic hardness test, scratch hardness test, rebound hardness test, abrasion and erosion test, and other contactless and non-destructive hardness tests. Hardness is testing method dependent and every tested result has to have a label identifying the testing method used. Among these hardness measurements, static indentation method is the most widely used method in the determination of hardness of steels. This method involves the formation of permanent indentation in the surface of the steel to be examined, the hardness being determined by the load and the size of the indentation formed.

Hardness test does not require much time and is very informative since there are relationships between hardness and other mechanical properties. For example, the hardness determined by static indentation test, H, is approximately proportional to the UTS under the indenter that no longer increases with increasing test force as:

$$\mathrm{UTS} \approx C \cdot H \tag{5.6}$$

where C is a constant. C depends on the material tested, the type of indenter, and other testing conditions, and has a value in the range from 3 to 3.55 for steel. Hardness test is extremely important in the evaluation of strength of a steel when tensile test cannot be performed due to the very small size of specimen [5].

5.3. Effect of Tungsten on Hardness

Figure 5.6 shows the variation of Vickers hardness with tungsten content of the forged 0W, 0.1W, 0.5W and 1W steels. The hardness was measured with a Vickers hardness tester using 2 kg load and 10 second dwelling time, and eight indentations for every specimen were made randomly on its surface.

The average value of these eight measurements is regarded as the hardness of the specimen tested. It can be seen that significant increase in the hardness has been caused even though only 0.1% tungsten is added. With continuously increasing the tungsten content from 0.1% to 1%, the hardness turns out gradual upward trend.

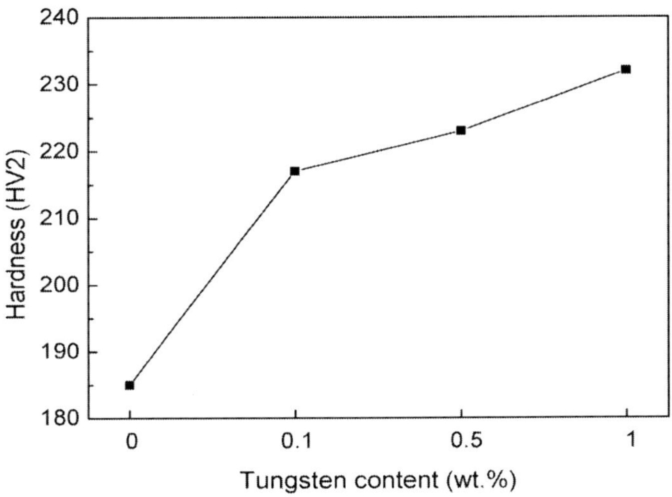

Figure 5.6. Variation of Vickers hardness with tungsten content.

5.4. Effect of Tungsten on Impact Toughness

Standard Charpy impact specimens were machined according to ASTM E23 [6]. Charpy V-notched specimens with cross section of 10 mm × 10 mm, length of 55 mm, notch angle of 45° and notch depth of 2 mm were employed to study the room temperature (RT) and low temperature (-50°C) impact fracture toughness on a Zwick/Roell impact tester. For each condition, three specimens were used to do the impact tests. Table 5.1 compares the absorbed impact energies of the forged steels tested at RT and -50°C. It is clear that the absorbed impact energy shows a decreasing trend with tungsten at both RT and -50°C. The values presented in Table 5.1 indicate that the impact toughness of all the steels at both RT and -50°C is very poor even though the absorbed impact energy at RT is higher than that at -50°C for the same steel.

Table 5.1. Absorbed impact energies of the forged steels tested at RT and -50°C

Temperature (°C)	0W	0.1W	0.5W	1W
RT	19.3 J	16.1 J	14.5 J	10.6 J
-50	2.46 J	2.11 J	1.77 J	1.41 J

Figure 5.7a and b show the fracture surfaces of the 0W steel after Charpy impact tests at RT and -50°C, respectively. It can be seen that the fracture surface tested at RT consists of cleavage facets, while it becomes a mixture of cleavage facets and shear cracks when tested at -50°C. Both fracture features exhibit the full brittle nature of fracture of the 0W steel at both RT and -50°C.

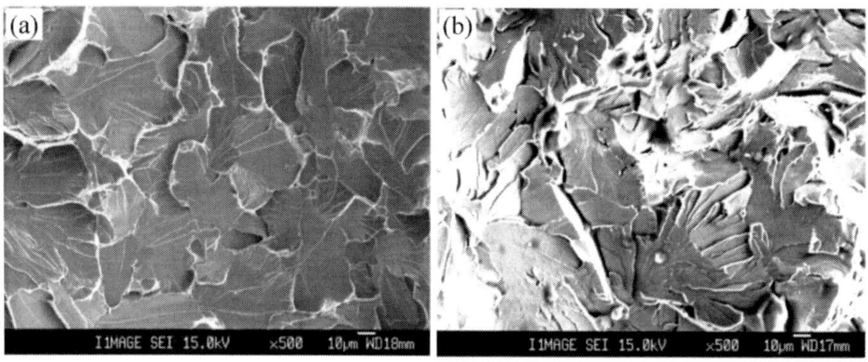

Figure 5.7. Fracture surfaces of the 0W steel after Charpy impact tests at (a) RT and (b) -50°C.

Brittle mode fractures are also expected for all other tungsten-containing microalloyed steels due to the fact of their very low level of absorbed impact energies, as shown in Table 5.1. The impact toughness of a steel is closely related to the microstructural constituents including phase composition, grain size and grain distribution etc. For enhancing the impact toughness of a steel, its microstructure should be modified through heat treatment and/or hot working to obtain an optimal microstructural constituent with high absorbed impact energy. A heat treatment technology will be introduced in Chapter 6, with the purpose of enhancing the impact toughness of these microalloyed steels with different tungsten additions.

5.5. Effect of Tungsten on Tensile Properties

5.5.1. Tensile Properties

Tests on tensile properties

Round bar tensile specimens were prepared according to ASTM E8 (gage length L_0 = 30 mm, diameter of the reduced section d_0 = 6 mm). Tensile tests were performed on an INSTRON 8862 tensile testing machine at RT and low temperatures (-50°C and -100°C) with strain rate of 5×10^{-3} s^{-1}. Three specimens were used in each condition for tensile tests. Low temperatures were controlled by liquid nitrogen through a Very Low Temperature Material Test Chamber. The temperature was monitored throughout each tensile test. Tensile specimen was maintained in the low temperature chamber for 5 min once the desired testing temperature was achieved. The deviation of testing temperatures was controlled within ±5°C. To avoid sliding from the clamp apparatus during tensile testing, both ends of the specimens were machined to be screw threaded, as shown in Figure 5.8 [7]. Specimens were attached firmly to special clamps with female screw thread on the INSTRON 8862 tensile testing machine. The resulting engineering stress-strain curves were analysed to determine the YS (0.2% proof stress) and UTS.

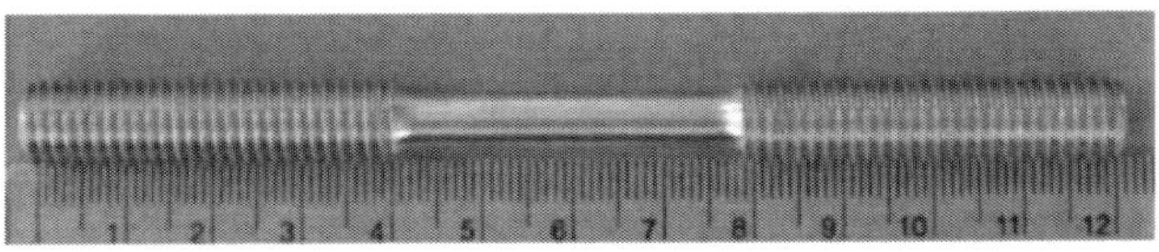

Figure 5.8. Dimensions and appearance of tensile specimen used [7].

Room/low-temperature tensile properties

Figures 5.9-5.11 present the typical tensile curves of the steels tested at RT, -50°C and -100°C, respectively [7]. The results show that the tensile behaviour at RT are similar to those at -50°C, i.e., all the engineering stress-strain curves without apparent yield points can be simply divided into three stages, such as elastic deformation, plastic deformation and fracture, as shown in Figures 5.9 and 5.10. With the temperature continuously decreasing from -50 to -100°C, the tensile behaviour is greatly changed.

As shown in Figure 5.11, the tensile curves at -100°C exhibit lower strain to failure as compared to those at RT and -50°C.

The tensile properties determined from the tensile curves are listed in Table 5.2 [7]. The relationships between tensile properties and tungsten content at various temperatures are presented in Figure 5.12 [7]. From Table 5.2 and Figure 5.12, it can be found that the addition of tungsten brings about increase in tensile strength at the expense of strain. Also, the YS and UTS of all the steels increase gradually with a decrease of the temperature from RT to -100°C (Figure 5.12a).

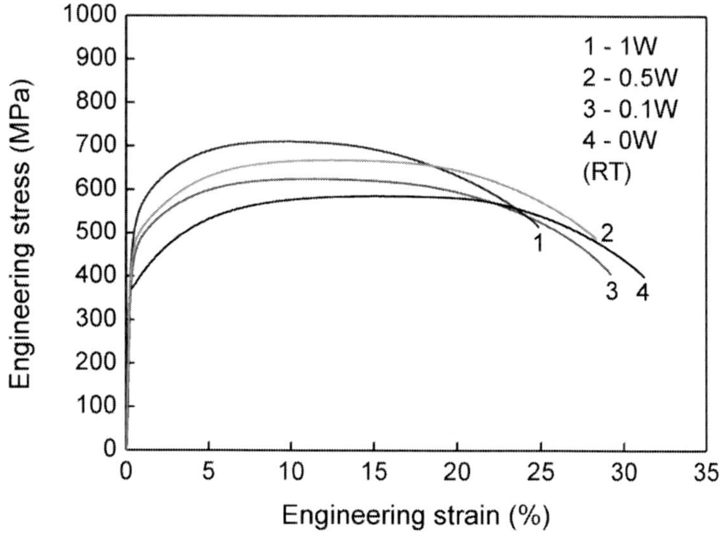

Figure 5.9. Typical tensile curves of the microalloyed steels tested at RT [7].

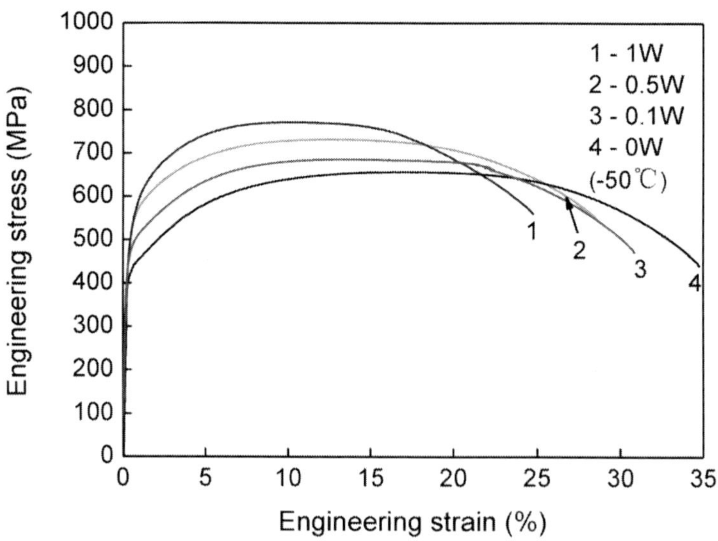

Figure 5.10. Typical tensile curves of the microalloyed steels tested at -50°C [7].

As demonstrated in Figure 5.12b, the strain to failure is higher when tested at -50°C than that at RT in the tungsten content range of 0 to 0.5%. When the tungsten content is increased to 1%, however, the fracture strain at -50°C is slightly lower than that at RT. When tested at -100°C, the fracture strain shows a very low value relative to that at both RT and -50°C, and the fracture strain decreases moderately in the tungsten content range of 0 to 0.5%, then drops sharply when the tungsten content is greater than 0.5% (Figure 5.12b).

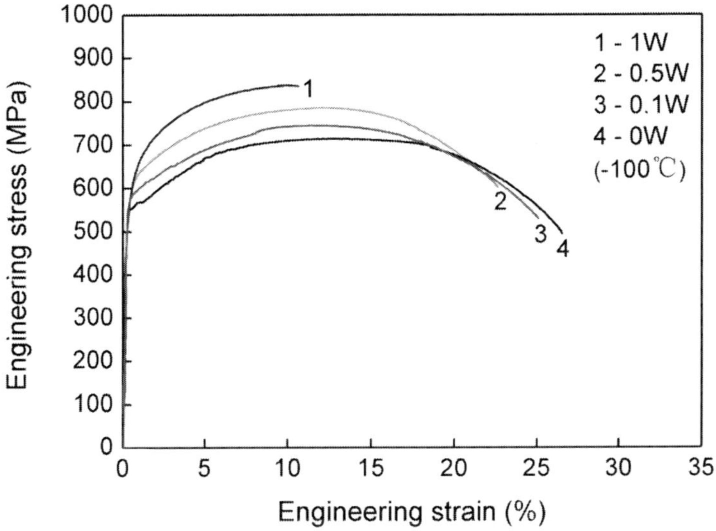

Figure 5.11. Typical tensile curves of the microalloyed steels tested at -100°C [7].

Table 5.2. Tensile properties of the microalloyed steels tested at various temperatures [7]

Steels	Temperature (°C)	YS (MPa)	UTS (MPa)	Engineering strain (%)
0W	RT	299.1	585.6	31.2
	-50	336.6	657.2	34.8
	-100	383.7	714.1	26.5
0.1W	RT	313.4	624.9	29.2
	-50	365.4	686.2	30.9
	-100	409.8	744.4	25.1
0.5W	RT	345.7	668.2	28.3
	-50	379.8	732.6	28.6
	-100	424.6	785.6	22.7
1W	RT	351.9	711.1	24.9
	-50	392.2	771.7	24.8
	-100	439.1	837.6	10.7

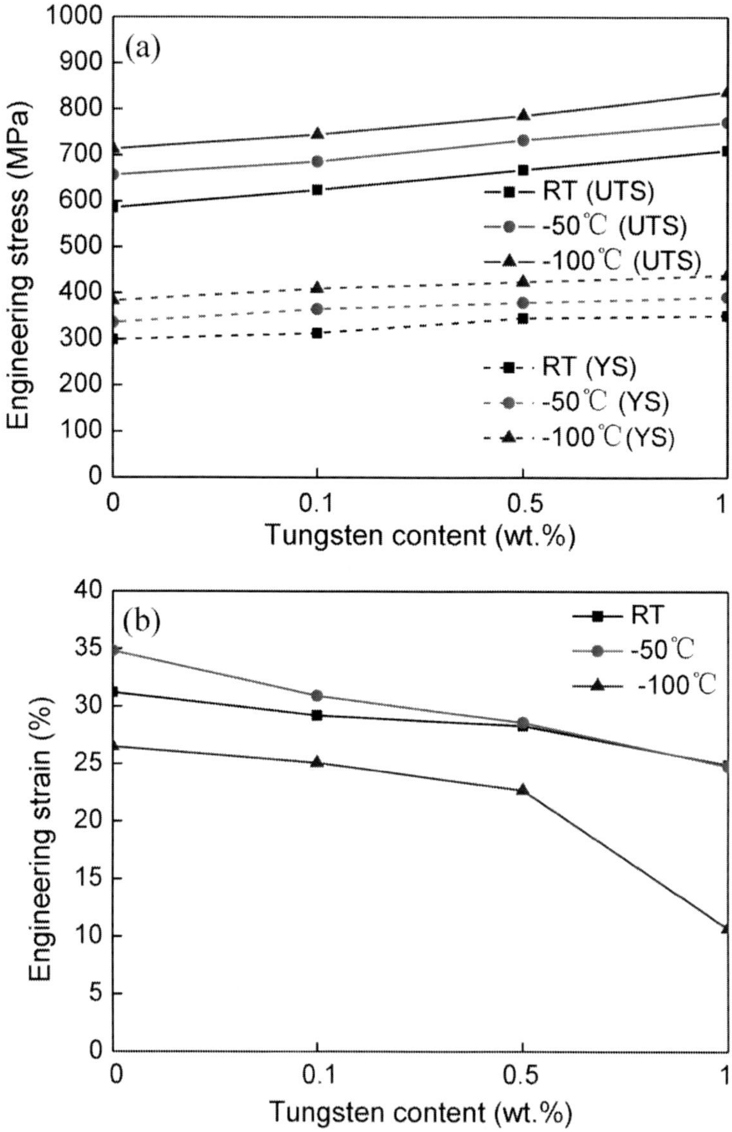

Figure 5.12. Relationships between tensile properties and tungsten content at various temperatures: (a) plots of stress vs. tungsten content, and (b) plots of strain vs. tungsten content [7].

5.5.2. Relationship between Tensile Properties and Temperature

Figure 5.13 shows plots of the logarithms of YS and UTS as functions of the inverse of the absolute testing temperature [7]. It is apparent that a linear relationship between $\ln\sigma$ and $1/T$ can be obtained as:

$$\ln\sigma = \ln A + n/T \qquad (5.7)$$

where σ is the stress, T is the absolute testing temperature, and A and n are constants. Then, the stress can be expressed as an Arrhenius type function of T as:

$$\sigma = A \cdot e^{n/T} \qquad (5.8)$$

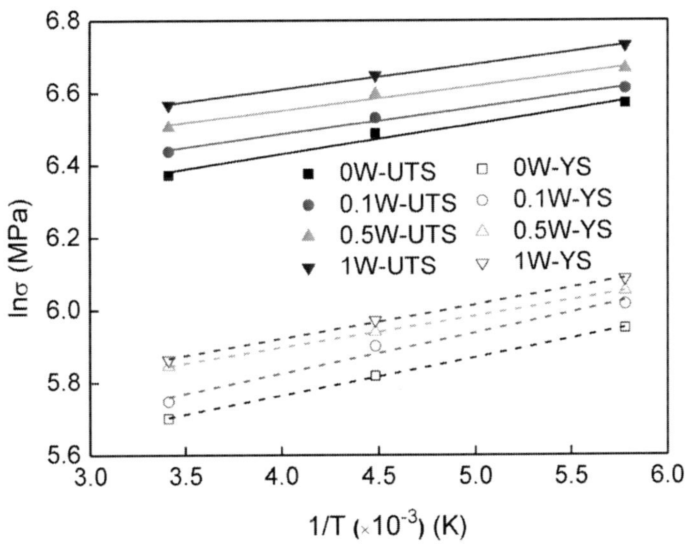

Figure 5.13. Plots of $\ln\sigma$ vs. $1/T$ of steels with different tungsten contents [7].

After linear regression from Figure 5.13, the values of $\ln A$ and n of the tested steels with different tungsten contents can be obtained, as presented in Table 5.3 [7]. Such relationships between σ and T can be used to estimate the YS and UTS of these steels at other testing temperatures.

Table 5.3. Values of $\ln A$ and n of the tested steels [7]

Stress	Constants	0W	0.1W	0.5W	1W
YS	$\ln A$	5.344	5.375	5.550	5.548
	n	0.105	0.112	0.087	0.093
UTS	$\ln A$	6.098	6.192	6.280	6.334
	n	0.083	0.074	0.068	0.069

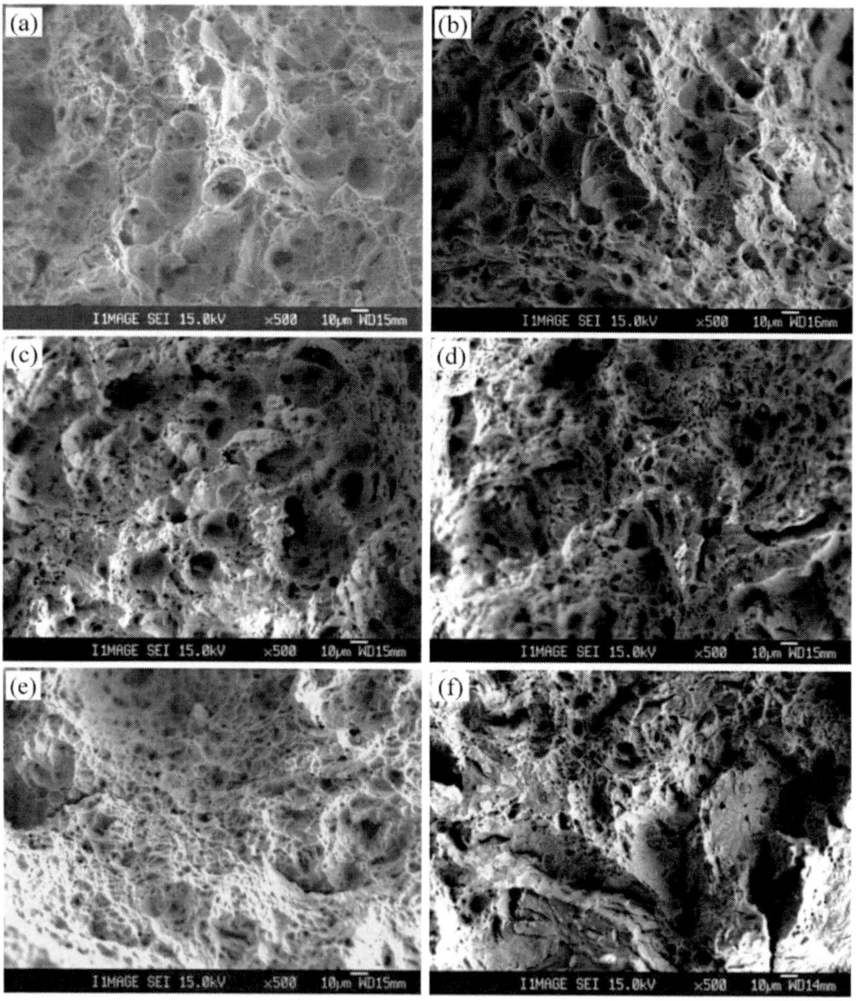

Figure 5.14. Fracture surfaces of the steels tested at (a,c,e) RT and (b,d,f) -50°C: (a,b) 0W, (c,d) 0.5W, and (e,f) 1W [7].

5.5.3. Fractography

Figure 5.14 shows the fracture surfaces of the steels tested at both RT and -50°C [7]. When tested at RT, the fracture surfaces of all the steels with different tungsten contents are composed of dimples, which indicate a characteristic of ductile fracture. Also, the size of dimples decreases with the increase of tungsten content, as shown in Figure 5.14a, c and e. As temperature decreases from RT to -50°C, the fracture mode of the 1W steel has been changed. As shown in Figure 5.14b and d, the fracture surfaces of the 0W and 0.5W steels still show ductile mode. However, when the tungsten content is increased to 1%, the fracture surface is mixed with cleavage facets and small dimples, as shown in Figure 5.14f. Moreover, with a decrease of the temperature from RT to -50°C, the size of the dimples shows a decreasing trend.

Figure 5.15 shows the macroscopic comparison of the fractured steels tested at -100°C. It can be seen that the necking phenomenon becomes less and less obvious with increasing the tungsten content, indicating a decreasing trend of strain with tungsten after tensile test.

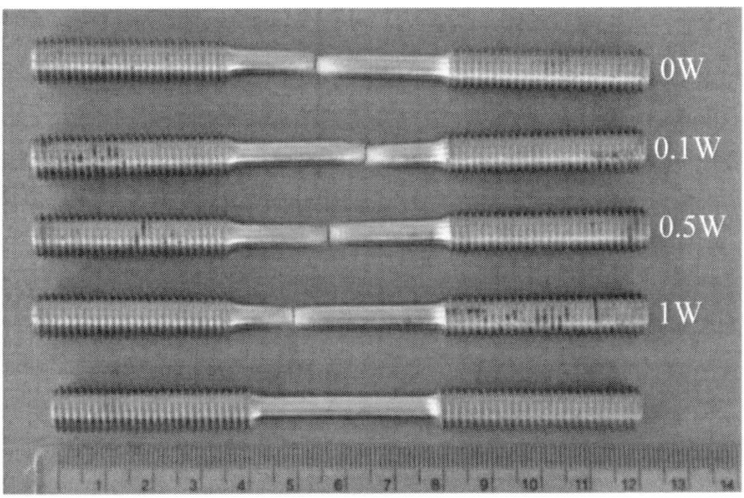

Figure 5.15. Macroscopic comparison of the fractured steels tested at -100°C.

Figure 5.16 shows the fracture surfaces of the steels tested at -100°C [7]. As illustrated in Figure 5.16a, c and e at low magnification, the reduced cross sectional area after fracture shows a decreasing trend with tungsten, which is in consistence with the observation from Figure 5.15. The fracture feature of the 0W steel still shows a ductile mode, as shown in Figure 5.16b. However, the fracture surface of the 0.5W steel consists of cleavage facets, small dimples and small quantities of cracks, as shown in Figure 5.16d. When the tungsten content is increased to 1%, the fracture surface becomes a mixture of cleavage facets and many short cracks, which means that full brittle fracture has occurred, as shown in Figure 5.16f. Comparing Figure 5.16f with Figure 5.14f, it follows that DBT of the 1W steel takes place when the temperature is decreased from -50 to -100°C.

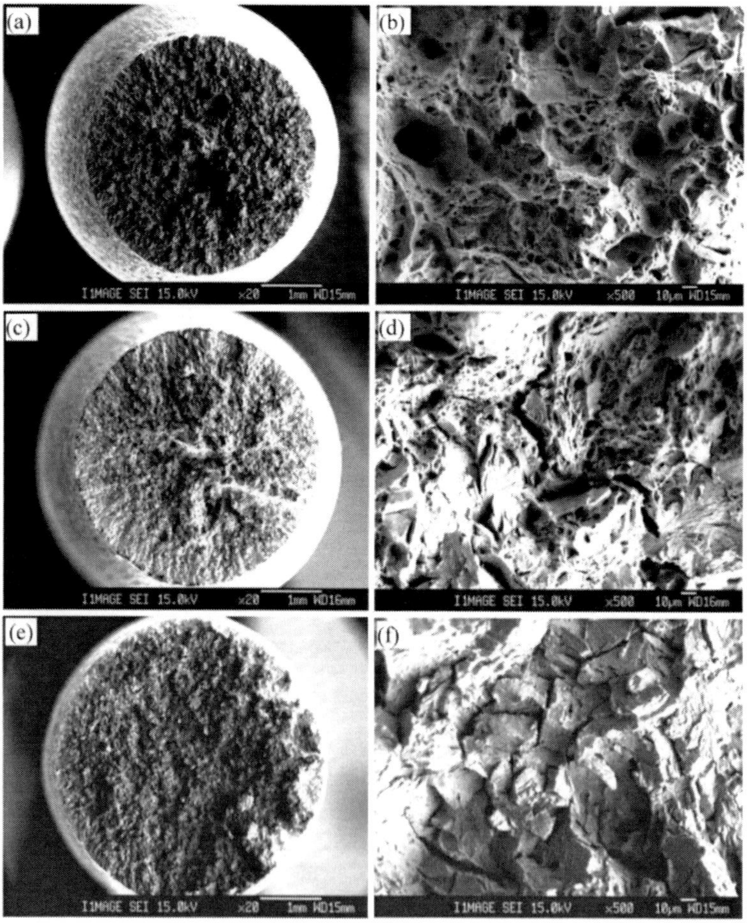

Figure 5.16. Fracture surfaces of the steels tested at -100°C: (a,b) 0W, (c,d) 0.5W, and (e,f) 1W [7].

Table 5.4 presents the characteristics of fracture surfaces of the steels tested at different temperatures [7]. As can be seen, steels with higher tungsten content exhibit higher temperature to DBT. Moreover, based on Table 5.4, it can be deduced that the 0.5W steel will be the next one to experience DBT as the temperature continuously decreases from -100°C to much lower. According to the above fracture surface examination and analysis, it can be concluded that the addition of tungsten has a positive effect on increasing the DBT temperature.

Table 5.4. Characteristics of fracture surfaces of the steels tested at different testing temperatures [7]

Temperature (°C)	0W	0.1W	0.5W	1W
RT	*D*	*D*	*D*	*D*
-50	*D*	*D*	*D*	*D* + *CF*
-100	*D*	*D*	*D* + *CF* + *C*	*CF* + *C*

D: dimples; *CF*: cleavage facets; and *C*: cracks.

5.6. Mechanism of Tungsten-Induced Variation of Mechanical Properties

When the tungsten content is increased from 0 to 0.5%, the amount of allotriomorphic ferrite increases gradually (see Figures 4.9 and 4.10 of Chapter 4), which subsequently leads to an increase in the volume fraction of AF. The shape change is always associated with the growth of AF, and such shape deformation causes plastic deformation in the adjacent austenite. As reported by Strangwood and Bhadeshia [8], the resulting dislocations are inherited by the AF as it grows, giving a dislocation density which is typically at 10^{14} m^{-2}, and which contribute some 145 MPa to its strength. Figure 5.17 contains TEM micrographs of dislocations in ferrite observed in the 0W and 0.1W steels [7,9]. It can be seen that large numbers of dislocations exist in the 0.1W steel relative to that in the 0W steel, and hence, higher hardness and strength of the 0.1W steel is expected. When the tungsten content is increased to 1%, allotriomorphic ferrite disappears, and the microstructure consists of mixed AF and BF, as shown in Figure 4.10c of Chapter 4. The TEM results obtained by Smith [10] revealed that the dislocation density of bainite ρ_d is about 4×10^{14} m^{-2}, which is higher than that of allotriomorphic ferrite obtained in the same steel with $\rho_d \approx 0.5 \times 10^{14}$ m^{-2}. In another supporting experiment, similar results for the dislocation densities of bainite and allotriomorphic ferrite were measured to be 1.7×10^{14} and 0.37×10^{14} m^{-2}, respectively [11]. It then follows that bainite contains more dislocations than allotriomorphic ferrite and hence improves the hardness and strength. As to the reasons for the increase in hardness and strength after tungsten addition, the following other causes are considered. One is the solid solution strengthening induced by tungsten addition, which results from the fact that the atomic size of tungsten is about 10% larger than that of Fe. The other is the hardening induced by the fine precipitates in tungsten-containing steels. With refinement of various precipitates, the barriers to dislocation motion increase with the increasing amount of interphase boundaries, leading to an increase in the hardness and strength in this way.

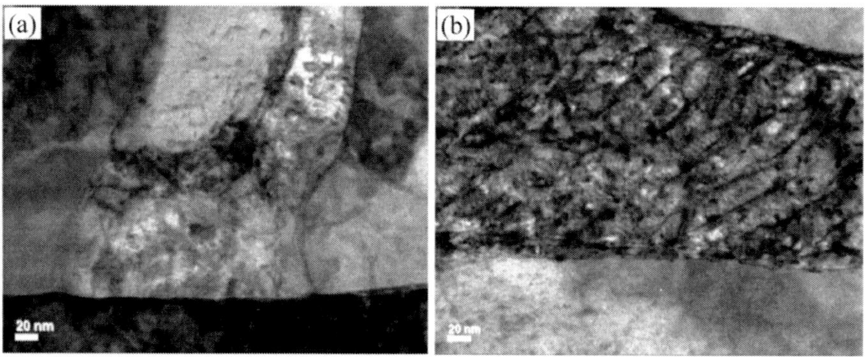

Figure 5.17. TEM micrographs of dislocations in ferrite observed in the (a) 0W and (b) 0.1W steels [7,9].

At low temperature, the flow stress of steels has been found to be strongly temperature dependent, as in other BCC and BCC-like materials [12-14]. As the temperature decreases, the increases in both YS and UTS are largely attributed to the larger effective stress needed to move the dislocations past short range obstacles at low temperatures as sufficient thermal energy for dislocation motion is lacking [15]. For BCC materials, it is well accepted that the low temperature plastic deformation behaviour is controlled by a Peierls lattice resistance mechanism [16-19]. In the Peierls mechanism, the dislocations experience the periodicity of the lattice as glide on their slip plane, which results in the presence of peaks and valleys of energy as a function of dislocation position. A moving dislocation causes bond angle distortions. BCC materials are strongly directional in their bonding, and the bond angle distortion necessary for dislocation motion in these materials is thus difficult to overcome. In BCC materials, the Peierls force is the primary obstacle to dislocation motion, even when lattice vibration energy is enhanced at high temperatures. BCC materials develop a directional bonding component at low temperatures. The movement of dislocations in BCC materials is thus strongly inhibited at low temperatures by the Peierls force. It is therefore expected that the strength of the studied steels, which are BCC structures, would exhibit an increasing trend with decreasing testing temperature. Because the matrix is highly dislocated, it would have large numbers of mobile dislocations even at low temperatures. The presence of these mobile dislocations would tend to limit the tendency towards cleavage and permit slip to take place [13]; there would therefore be resistance to necking in tensile tests and good strain values, would be exhibited, particularly at -50°C. Since the dislocation motion is retarded with the refinement of precipitates after tungsten addition, the increment of strain at -50°C relative to that at RT shows a decreasing trend with tungsten (Figure 5.12b). When the testing temperature is decreased to -100°C, the sharp decrease in strain is probably caused by the considerable reduction in mobile dislocations at such low temperature.

5.7. Summary

In this chapter, mechanical tests including tensile test, impact test and hardness test are introduced to evaluate the mechanical properties of microalloyed steels. The effects of tungsten addition on the hardness, room/low-temperature impact toughness and room/low-temperature tensile properties of several microalloyed forging steels are analysed.

All the microalloyed forging steels exhibit a low level of impact toughness, and the absorbed impact energy shows a decreasing trend with tungsten. Tungsten plays an important role in increasing the hardness and tensile strength at the expense of strain. With a decrease of the testing temperature, both YS and UTS show an increasing trend. The tensile strength is strongly temperature dependent, and the relationship between tensile strength and temperature follows an Arrhenius type equation as $\sigma = Ae^{n/T}$.

When the tungsten content is equal to or less than 0.5%, the fracture strain at -50°C is higher than that at room temperature. At -100°C, the fracture strain of all the tested specimens showed is very low relative to that at RT and -50°C. With a decrease of temperature from -50 to -100°C, DBT of the specimen with

1% tungsten occurs. The addition of tungsten shows a positive effect on increasing the DBT temperature.

It is suggested that the mechanical properties of microalloyed forging steels are closely related to the tungsten content, and the addition of 0.5% tungsten yields the best combination of excellent room/low-temperature tensile strength and ductility.

REFERENCES

[1] ASTM E8/E8M, Standard Test Methods for Tension Testing of Metallic Materials, *ASTM International*, 2013.

[2] M. Lange, *Natural Laws in Scientific Practice*, Oxford University Press, 2000.

[3] J.M. Holt (Ed.), Charpy Impact Test: Factors and Variables, *ASTM International*, 1990.

[4] A. Nayar, *Testing of Metals,* Tata McGraw-Hill Education, 2005.

[5] A. Nayar, *The Steel Handbook*, Tata McGraw-Hill Education, 2000.

[6] ASTM E23, Standard Test Methods for Notched Bar Impact Testing of Metallic Materials, *ASTM International*, 2007.

[7] J. Zhao, T. Lee, J.H. Lee, Z. Jiang, C.S. Lee, Effects of tungsten addition on the microstructure and mechanical properties of microalloyed forging steels, *Metallurgical and Materials Transactions A*, 44A (2013) 3511-3523.

[8] M. Strangwood, H.K.D.H. Bhadeshia, The mechanism of acicular ferrite formation in steel weld deposits, in: S.A. David (Ed.), *Advances in Welding Science and Technology*, ASM International, Metals Park, Ohio, 1987.

[9] J. Zhao, Y.W. Kim, J.H. Lee, J.M. Lee, H.S. Chang, C.S. Lee, Effect of tungsten addition on the mechanical properties and corrosion resistance of S355NL forging steel, *Metals and Materials International*, 18 (2012) 217-223.

[10] G.M. Smith, *The Microstructure and Yielding Behaviour of Some Ti Steels*, University of Cambridge, 1984.

[11] M.K. Graf, H.G. Hillenbrand, P.A. Peters, Accelerated cooling of plate for high-strength large-diameter pipe, in: P.D. Southwick (Ed.), *Accelerated Cooling of Steel, TMS-AIME*, 1985.

[12] W.A. Spitzig, A.S. Keh, Orientation dependence of the strain-rate sensitivity and thermally activated flow in iron single crystals, *Acta Metallurgica*, 18 (1970) 1021-1033.

[13] M.R. Krishnadev, R. Ghosh, Low-temperature mechanical behavior of an "acicular" ferrite HSLA (high strength-low alloy) line pipe steel, *Metallurgical Transactions A*, 10A (1979) 1941-1944.

[14] S. Naamane, G. Monnet, B. Devincre, Low temperature deformation in iron studied with dislocation dynamics simulations, *International Journal of Plasticity*, 26 (2010) 84-92.

[15] W.A. Spitzig, A.S. Keh, The role of internal and effective stresses in the plastic flow of iron single crystals, *Metallurgical Transactions*, 1 (1970) 3325-3331.

[16] C. Keller, M.M. Margulies, Z. Hadjem-Hamouche, I. Guillot, Influence of the temperature on the tensile behaviour of a modified 9Cr–1Mo T91 martensitic steel, *Materials Science and Engineering A*, 527 (2010) 6758-6764.

[17] S. Vaynman, M.E. Fine, S. Lee, H.D. Espinosa, Effect of strain rate and temperature on mechanical properties and fracture mode of high strength precipitation hardened ferritic steels, *Scripta Materialia*, 55 (2006) 351-354.

[18] P. Spätig, G.R. Odette, G.E. Lucas, Low temperature yield properties of two 7–9Cr ferritic/martensitic steels, *Journal of Nuclear Materials*, 275 (1999) 324-331.

[19] B.A. Hands, H.M. Rosenberg, The low temperature yielding and flow of potassium and its alloys, *Acta Metallurgica*, 17 (1969) 455-461.

Chapter 6

HEAT TREATMENT OF MICROALLOYED STEELS WITH AND WITHOUT TUNGSTEN ADDITIONS

ABSTRACT

Heat treatment involves heating a steel to an elevated temperature for sufficient time followed by controlled cooling with the purpose of obtaining particular microstructures and combinations of properties. Heat treatment of steels is an important operation in the final fabrication process of many engineering products prior to being put into service. In this chapter, a brief introduction of the purpose of heat treatment of steels and the parameters involved in heat treatment is first given. Then the concepts of treatment processes including annealing, normalising, hardening and tempering that are commonly used in practice for steels especially for hypo-eutectoid steels are presented. Next, carbon segregation in a microalloyed cast steel is analysed, and heat treatment processes are designed with the purpose of acquiring the optimal heat treatment conditions correlated with the best combination of excellent toughness and tensile properties. Several microalloyed forging steels with different tungsten additions are then heat treated in order to understand the effects of tungsten content, heat treatment temperature and cooling method on the microstructure, impact toughness and tensile properties, and to obtain the optimal heat treatment conditions of these steels. Finally, a heat treatment technology is successfully applied to a heavy sectional microalloyed steel to obtain uniformly distributed chemical compositions, microstructure and mechanical properties.

6.1. Purpose of Heat Treatment of Steels

The fundamental purpose of heat treatment of steels is to modify their structure and physical (sometimes chemical) properties through controlled heating and cooling in order to increase the service life of steel products by increasing the strength or ductility, or prepare the steels for improved manufacturability. Most of the engineering properties, especially mechanical properties, of steels are related to their atomic structure, crystal structure and microstructure. The same steel can have a very wide range of mechanical properties by adjusting the nature, distribution and quantities of these micro constituents when the steel is subjected to different heat treatments. Heat treatment process involves heating steel to a specific predetermined temperature with a certain rate, then holding the steel at this temperature for a required time, and finally cooling it to room temperature under a specific cooling rate. All these heat treatment operations are carried out in the solid-state of steels. Vital factors involved in a heat treatment in determining the outcomes are the temperature to which the steel is heated and the rate at which it is cooled down. Therefore, heat treatment of steels may also be defined as a sequence of heating and cooling operations designed in the solid-state to obtain the desired combination of properties.

Heat treatment of steels is a very important operation in the final fabrication of many engineering steel components. It is of significance in making the steel parts and tools better suited, structurally and physically, for sophisticated applications. Heat treatment of steels still plays a very important role in the development of high performance steels, such as advanced high strength steels (AHSS), when science and technology are advancing very rapidly today.

6.2. Heat Treatment Parameters of Steels

Heating rate, heat treatment temperature, holding time and rate of cooling are four key heat treatment parameters that significantly affect the outcomes of the steels to be heat treated. The magnitude of these parameters required depends on the chemical compositions, size and shape of the object and the desired properties of the steel after heat treatment.

6.2.1. Heating Method

Two kinds of heating methods are generally applied in practical heat treatment of steels. One method involves putting the steel into a heat treatment furnace at room temperature firstly, and then heating up the furnace to the predetermined temperature. In the other method, steel to be heat treated is put into the furnace that has been heated up to the target temperature in advance. The first heating method is extensively used in experimental research and real production lines of steels; whereas the second one is usually used when the size of heat treated steel is small. In some cases, both heating methods are applied in the combination of the use of two furnaces in order to reduce the oxidation problem occurred during heat treatment. That is, in one furnace, the steel is heated up to a temperature at which there is no appreciable oxidation. Then the steel is transferred to the second furnace in which the target temperature has been reached. Such operations can reduce the total time of holding at high temperature and thus can reduce the occurrence of oxidation.

6.2.2. Heating Rate

The heating rate of a furnace can be pre-set as a constant through a temperature control panel available with the furnace. In practice, however, the rate of heating is not a constant when a steel object is in the furnace. For such a condition, the average rate of heating, which is the total increment in temperature divided by the total time taken, is normally used.

The size and shape, and the thermal conductivity of the steel to be heat treated affect significantly the heating rate that should be applied. Generally, the larger the size of the steel, the lower shall be the heating rate. Distortions will be caused during the heat-up of steels, including (i) shape change due to relief of residual stress, (ii) shape change due to thermal stresses caused plastic flow, and (iii) volume change due to phase changes on heating (e.g., the pearlite-to-austenite phase change causes approximately a 4% contraction on heating). A low heating rate can reduce the development of these distortions caused by the thermal gradient and phase changes, thus ensuring temperature homogeneity of the steel and reducing the holding time required at the heat treatment temperature. Steels with complicated shape and sharp edge/corner should be heated slowly. A steel that has variable sections should also be heated at a low rate. For the steels with relatively low thermal conductivity, such as austenitic stainless steel grades, a low heating rate should be applied.

Moreover, for the steels that are prone to oxidation at high heat treatment temperatures, faster heating rates are recommended within the high temperature range.

6.2.3. Heat Treatment Temperature

Heat treatment temperature is governed by the chemical compositions of the steel to be heat treated and the properties required after heat treatment. Some critical temperatures include A_1 and A_3 are of vital importance and should be known before conducting the heat treatment of steel. Commonly, a CCT diagram, in which the critical points for phase transformations are clearly identified, can be constructed as a valuable reference before the design of heat treatment processes. For example, for normalising heat treatment of microalloyed steels, they should be heated up above the critical temperature A_{c3} which can be directly obtained from the relevant CCT diagram.

6.2.4. Holding Time

After heating up to the heat treatment temperature, the steel is generally held at this temperature for 1 h per 25 mm thickness. For a steel object with variable section thicknesses, the holding time is determined on the basis of the thickest section. For larger sections, holding time is increased at the rate of 30 minutes for each additional 25 mm of thickness to ensure both the temperature and microstructure are uniformly distributed. A longer holding time is generally applied if the steel is heated up with a relatively high heating rate. The holding time should be reduced with the increase of heat treatment temperature. In practice, the holding time may be adjusted based on the nature of the steel to be heat treated, the heat treatment temperature predetermined and the dimensions of the heat treated steel object.

6.2.5. Cooling Rate

The cooling rate of a heat treated steel object is determined by the microstructure required after heat treatment. In most cases, it is possible to acquire different microstructural combinations of steel through control of phase transformation behaviours by changing the cooling rates from heat treatment temperatures. With the aid of CCT diagrams of microalloyed steels, for example, microstructures of “ferrite + pearlite”, “ferrite + pearlite + bainite”, “ferrite + bainite”, “bainite + martensite” and “full martensite” can be obtained by changing the cooling rate from 0.1 to 120°C/s, as already discussed in Chapter 2. For the heat treatment applied in real production lines of steels, however, accurate control of cooling rate is not practical, and sometimes even impossible. Three cooling modes including quenching, air cooling and furnace cooling are extensively applied in scientific research and steel industries. In certain cases, cooling combined with heating may vary from a most simple form of heating in a closed furnace and cooling in the one of the aforementioned cooling medium to a more complex form which requires preheating, holding in a controlled atmosphere and two-step cooling. It must be noted that the thicker the section of the heat treated steel object, the slower should be the rate of cooling regardless of the method of cooling used except in such operations as induction hardening to be discussed later.

6.3. Classifications of Heat Treatment of Steels

Depending on the method of heating, the heating rate, the furnace atmosphere, the heat treatment temperature, the cooling medium applied and the properties desired, heat treatment processes of steels especially hypo-eutectoid steels that are commonly used in practice can be classified into the following basic types.

6.3.1. Annealing

Annealing is a treatment process involving heating the steel to a predetermined temperature, then holding at this temperature for a certain time, and finally cooling at a slow rate. Slow cooling is ordinarily achieved in a closed furnish by switching-off the supply, although it may be taken out and cooled in other medium such as mica, lime or ashes that can ensure a slow rate of cooling. The slow cooling process benefits the formation of spheroidal carbides and coarse lamellar pearlite, which are “soft” and result in the softening of steel. In annealing treatment, the cooling rate varies from 30 to 50°C/h. A lower cooling rate should be applied for microalloyed steels as compared to plain carbon steels.

The purposes of an annealing treatment of steel are to:

(1) Relieve internal stresses caused during solidification, machining, forging, rolling or welding.
(2) Reduce hardness.
(3) Enhance machinability.
(4) Improve or restore ductility and toughness to facilitate further cold working.
(5) Eliminate chemical non-uniformity.
(6) Reduce gas content in steel.
(7) Increase dimensional stability.
(8) Produce a definite microstructure having desired mechanical, physical or other properties.

Depending on the specific purpose in practical applications, annealing of a microalloyed steel can by classified into various types such as diffusion annealing, full annealing, isothermal annealing, incomplete annealing, recrystallisation annealing and process annealing, etc.

- **Diffusion annealing.** This process, also known as homogenising annealing, is performed to remove chemical segregation in ingots and castings by diffusion. In diffusion annealing treatment, steel is heated 150-300°C above the A_{c3} temperature. For carbon steels, the temperature is usually within 1100-1200°C; for alloyed steels, however, a higher temperature (commonly 1200-1300°C) should be applied. The diffusion annealing is usually carried out for more than 10 h. The holding time should not be too long since serious oxidation problem may happen during such a long holding time at the very high temperature. Heating to such a temperature results in considerable coarsening of austenite grains which cause the formation of coarse pearlite structure on cooling. The coarse grain structure can be eliminated by subsequent hot working of the ingots, or by employing a second heat treatment such as full annealing or normalising of the castings.
- **Full annealing.** Generally, the term "annealing" means full annealing unless otherwise specified. In full annealing, steel is heated up above the A_{c3} temperature and held at this temperature for a desired length of time, and then cooled very slowly. The temperature range normally used for full annealing is 30-50°C above the A_{c3} of hypo-eutectoid steels. This temperature produces maximum grain refinement with

homogeneous, reoriented and recrystallised stress-free austenite grains. Heating up too much above A_{c3} results in coarse austenite grains which impair the properties of the steel after annealing treatment. Higher annealing temperatures for microalloyed steels, especially those containing strong carbide forming elements such as vanadium, niobium and titanium, are recommended as compared to plain carbon steels and other alloy steels. The rate of cooling in full annealing must be slow enough to allow the formation of equilibrium structures as predicted by an equilibrium diagram at very low degrees of supercooling.

- **Isothermal annealing.** This treatment consists of austenitising the steel at the full annealing temperature for some time and then cooling rapidly to the appropriate temperature below A_{r1}. Fast cooling can be achieved by transferring steel to another furnace maintained at the desired temperature. The isothermal temperature is generally 10-30°C below A_{r1} that within the pearlite transition temperature. The steel is held at this temperature for a predetermined time till all the austenite gets transformed to pearlite. After the austenite-to-pearlite transformation is complete, steel is commonly cooled in air. For all practical purposes, the microstructure transformed after isothermal annealing is equivalent to that obtained by full annealing. Isothermal annealing is usually substituted for full annealing in order to cut short the annealing treatment cycle to a considerable extent.
- **Incomplete annealing.** Incomplete annealing is also referred to as partial annealing or inter-critical annealing. This treatment involves heating up the steel to be an appropriate temperature between A_{c1} and A_{c3} for some time, then cooling the steel slowly. In practice, the temperature applied is about 10-30°C above A_{c1} for hypo-eutectoid steels. The purposes of incomplete annealing of steel are to refine the grain size, homogenise the microstructure, relieve the internal stress, reduce the hardness and improve the machinability. Note that steels with coarse microstructure of ferrite and pearlite or with Widmanstätten structure are not suitable for this treatment because only partial transformation takes place. The major advantage of incomplete annealing over full annealing lies that it is less expensive than full annealing which involves higher annealing temperatures.
- **Recrystallisation annealing.** Recrystallisation annealing is an annealing process applied to cold-worked steels to obtain nucleation and growth of new grains without phase changes. In this treatment,

steel is heated above the recrystallisation temperature, then held at this temperature and cooled thereafter. After recrystallisation annealing, the cold-deformed grains are transformed to homogeneous and equiaxed grains which can decrease hardness or strength, remove residual stress, and increase ductility. In practice, the temperature applied for recrystallisation annealing is usually below and close to the A_{c1} temperature. The temperature for recrystallisation depends on the chemical compositions of the steel, the degree of deformation after cold working, the holding time and the initial grain size. The larger the degree of deformation, the lower the recrystallisation temperature. Since the effect of temperature is more predominant than that of holding time on recrystallisation, an increase of temperature is always recommended than that of holding time even though recrystallisation can be achieved at lower temperatures by increasing the holding time. Higher recrystallisation annealing temperatures are required for alloy steels and high carbon steels relative to other hypo-eutectoid steel grades.

- **Process annealing.** Process annealing, also called intermediate annealing, subcritical annealing, in-process annealing, or stress-relief annealing, is usually applied to cold-worked steels to restore their ductility in order to facilitate further cold working. Ductility is an important performance index that should be considered in shaping and creating a more refined piece of work through processes such as forging, rolling, stamping, drawing and extruding. Process annealing consists of heating the steel to a temperature below A_{c1}, holding at this temperature for sufficient time, and then cooling. The cooling rate should not be too fast because new internal stress may be caused during the fast cooling process. For practical purposes, the steel can be cooled slowly in the furnace to 500°C (300°C for heavy sectional steel), then taken out and cooled to room temperature in air. This process differs from recystallisation annealing in the sense that complete recrystallisation of the cold-worked steel may or may not take place in this treatment.

6.3.2. Normalising

Normalising is a heat treatment technique used to refine grains, homogenise microstructure, relieve internal stresses, enhance machinability, and improve mechanical properties of structural low alloy and carbon steels, etc. Normalising consists of heating the steel above the critical temperature A_{c3}, holding at this temperature for proper time until the temperature throughout the section is uniform and the structure is completely austenitised, and then cooling to room temperature in air. Typically, hypo-eutectoid steel is heated up to a temperature of approximately 50-60°C above the A_{c3}. Holding time is determined by the chemical compositions, and the shape and size of the steel object being heat treated. Exceeding the recommended temperature range or holding at the normalising temperature for a longer time may result in abnormal austenite grain growth.

Normalising differs from full annealing in that the cooling rate is accelerated by allowing the steel being cooled in air, and normalising temperature is somewhat higher than full annealing. The temperature ranges normally used for normalising and some types of annealing treatment processes are schematically shown in Figure 6.1.

The grain size in normalised structure is finer than that in annealed structure, and the normalised steels are generally stronger and harder than full annealed steels.

After normalising, the resultant microstructure is generally composed of fine pearlite and ferrite in hypo-eutectoid steels depending on their chemical compositions. In some cases, e.g., for the tungsten-containing microalloyed steels, bainite may form in the normalised microstructure. For such steels, slower cooling rates are required in order to obtain a refined pearlite plus ferrite microstructure.

Refinement of grain size is one of the most important targets of normalising. Almost all castings, hot forged and rolled steels that possess coarse grains due to high temperature involved in these operations are commonly subjected to normalising treatment for grain refinement. In some instances, double normalising is applied to steel castings and forgings with heavy sections in order to optimise microstructure and improve mechanical properties.

The second normalising treatment is carried out at the conventional normalising temperature, whereas the temperature for the first normalising treatment should be higher than that used for conventional normalising. Practically, the first normalising treatment is carried out at about 50-100°C higher than that used for the second normalising.

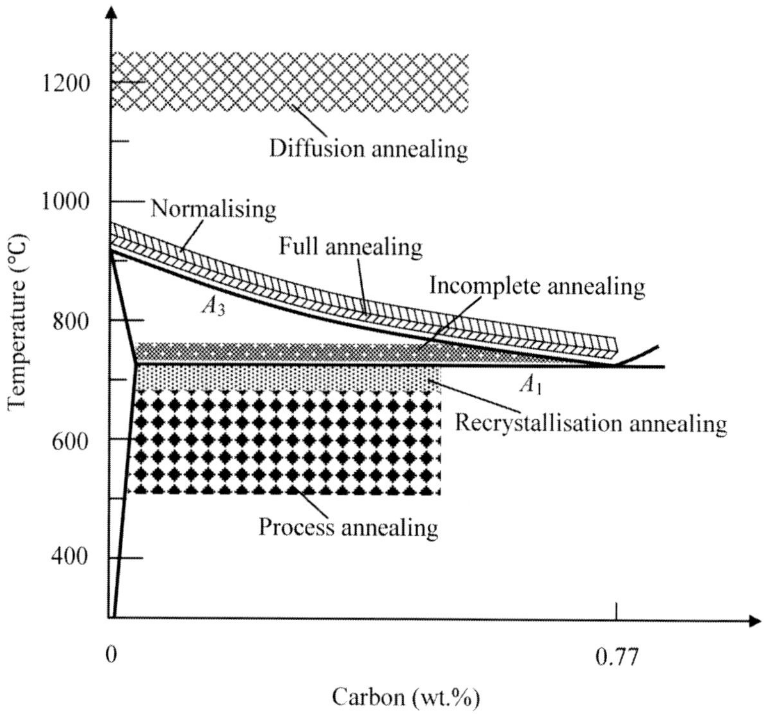

Figure 6.1. Schematic illustration of the temperature ranges for normalising and some annealing treatment processes.

6.3.3. Hardening

"Hardening", unless otherwise specified, refers to the heat treatment operation which consists of heating up to a predetermined hardening temperature (austenitising temperature), holding at this temperature until the temperature throughout the section is uniform and the microstructure is completely austenitised, and then cooling rapidly in a suitable medium such as water, oil or salt baths. Other hardening processes including surface hardening, induction hardening, case hardening, flame hardening, carburising, nitriding and carbonitriding are beyond the scope of this book. The hardening temperature depends on the chemical compositions of the steel being treated. Hypo-eutectoid steels are usually heated to about 30-50°C above the A_{c3} temperature. The term "cooling rapidly" means the rate of cooling is fast enough to suppress the austenite-to-ferrite, austenite-to-pearlite and/or austenite-to-bainite transformations with the purpose of obtaining a hard micro-constituent, i.e., martensite, in the resultant microstructure. The hardness finally achieved depends on the chemical compositions of the steel, the size and shape of the steel object, the hardening cycle (heating rate, hardening temperature, holding time and cooling rate), the homogeneity and grain size of austenite, the cooling medium and the surface condition of the treated steel.

One of the most important objectives of hardening is to develop high hardness and to increase the wear resistance of steels. The higher the hardness, the higher the wear resistance of the products will be. Hardening treatment is frequently used for spindles, gears, shafts, tools and dies, etc., which need high hardness. It is also useful for structural steels to have required mechanical properties. However, because of rapid cooling, high internal stresses are developed in the hardened steel. Also the as-hardened steel, while very hard, is too brittle to be useful for most applications. Therefore, the steel is rarely used in the as-hardened condition. Hardening is always followed by tempering in order to have an optimal combination of hardness, strength, ductility, toughness and elasticity.

6.3.4. Tempering

Tempering is a heat treatment of reheating hardened or normalised steels to a temperature below the A_{c1} temperature for sufficient time, and then cooling at a suitable rate. Most applications require that as-hardened steel parts be tempered to decrease the internal tresses and reduce the degree of brittleness. Martensite is a metastable phase. On tempering above room temperature, it decomposes to the more stable phases, ferrite plus carbides. After tempering, the hardness and strength of martensite falls, whereas the ductility and impact toughness are generally improved.

The nature of processes occurring during the tempering of martensite essentially depends on the tempering temperature and may be divided into the following three distinct stages.

- **Low temperature tempering (150-250°C).** Tempering in this temperature range results in the formation of a low carbon martensite and a carbide by transformation from high carbon martensite. The carbide precipitated from the high carbon martensite is not cementite (Fe_3C). This carbide, which is known as epsilon carbide ($Fe_{2.4}C$), has a hexagonal closed packed structure. Epsilon is metastable with respect to cementite. It forms at low temperature range where martensite is metastable. Hence the carbide size is very fine. Tempering in this low temperature range produces a modest increase in toughness that is adequate for applications that require high strength and fatigue resistance or loading is primarily compressive as in bearings and gears. Toughness also improves but to a lesser degree as compared to strength. Excellent wear resistance and reduced internal stresses are the characters of hardened steel tempered in such low temperature range. Low temperature treatment is useful for high carbon and low alloy steels used for manufacturing cutting and measuring tools.

- **Medium temperature tempering (350-500°C).** In this temperature range, retained austenite transforms to bainite which is composed of ferrite and epsilon carbide. When tempering is carried out in this medium temperature range, both ductility and toughness increase with a corresponding decrease of hardness and strength. The steel being tempered in this temperature range develops maximum elastic properties and the treatment is very suitable for coil and laminated springs.
- **High temperature tempering (500-680°C).** When tempering within such high temperature range, martensite changes to ferrite by losing its carbon. At the same time, carbon released combines with epsilon carbide to form cementite. Thus, high temperature tempering induces the formation of ferrite/cementite mixture. The steel treated at high tempering temperature is free from internal stresses, and has enhanced strength and toughness as compared to the annealed or normalised steel. This process applies to medium carbon and medium carbon low alloy steel products such as rods, shafts and gears, etc.

It should be mentioned that the above tempering temperature ranges specified may vary to a certain extent since they depend on the chemical compositions of steel, rate of heating, time of holding at the tempering temperature and type of measurement and sensitivity.

Tempering between 260°C and 370°C may cause tempered martensite embrittlement in hardened low alloy and medium carbon steels [1]. As a result, such tempering range is generally avoided in commercial practice. For applications requiring high toughness and less strength, tempering is done at temperatures above approximately 370°C. When such high temperatures are used, the steel object may be quenched from the tempering temperature to minimise the time in the embrittlement range. Another type of embrittlement, temper embrittlement, may develop in hardened carbon and alloy steels between 375°C and 575°C after tempering for relatively long time in or cooling slowly through this temperature range [2]. If tempering is carried out at a temperature higher than this critical range, the steel should be cooled rapidly through this temperature range, for example by quenching in oil, in order to avoid temper embrittlement. Tempered martensite embrittlement and temper embrittlement are both reversible. They can be eliminated by heating steel above the embrittlement range for only a few minutes followed by rapid cooling.

At any selected tempering temperature, the hardness drops rapidly at first, gradually decreasing more slowly as the holding time is prolonged. Short tempering periods are generally undesirable and should be avoided. The time for holding at tempering temperature may vary from 30 minutes to several hours. A longer tempering time at a given temperature increases the extent of tempering. Good practice requires a holding time preferably 1-2 h at tempering temperature for any hardened steel.

6.4. Heat Treatment of Microalloyed Cast Steel

6.4.1. Carbon Segregation

Microalloyed cast steel products are produced by a solidification process. The microstructure can be finely tuned, such as the grain structure, phase transformations and precipitation. However, cast steel generally consists of coarse and non-uniform microstructure which may impair greatly the mechanical properties. In addition, elemental segregation defect is also intimately linked to solidification. Elemental segregation (e.g., carbon) is detrimental to casting quality, and defective final product may be produced. During the production of cast steel products, this is of serious concern to foundries, and often results in scrapped casting or expensive repairs.

Figure 6.2 shows the OM structure of the 0W cast steel, which was manufactured by vacuum induction melting method [3]. As demonstrated, the grain size of the cast steel is coarse, and dendrite has been caused during casting process. Such microstructural non-homogeneity will be undoubtedly harmful to the mechanical properties especially toughness.

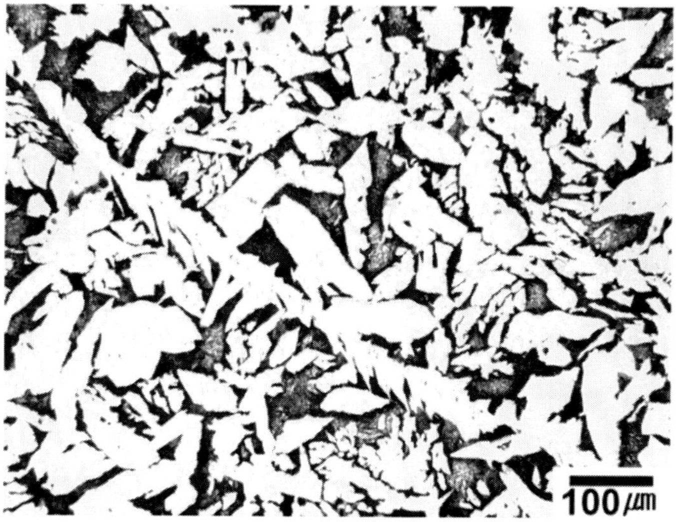

Figure 6.2. OM structure of the 0W cast steel indicating dendrite has been caused during casting process [3].

Figure 6.3 presents the result of EDS line-scanning across a dendrite arm boundary of the 0W cast steel [3]. High carbon peak is found at dendrite arm boundary, indicating carbon segregation was caused during solidification. In order to investigate the distribution of carbon in the cast steel, spot analysis, where the electron beam is focused into a fine spot on the site of interest, was performed to quantitatively measure the concentration of carbon along dendrite arm and grain boundaries, as discussed below.

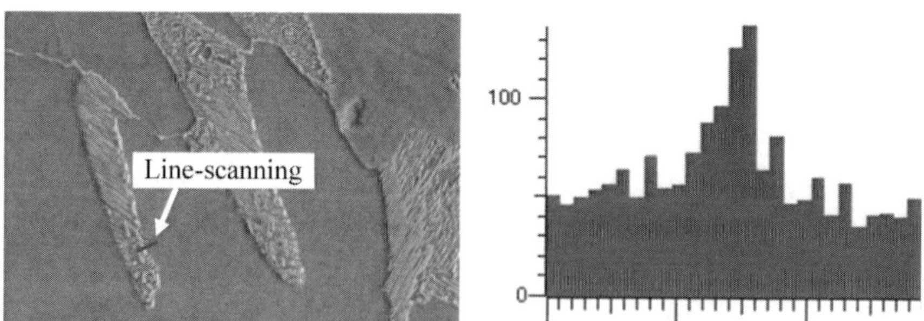

Figure 6.3. Line-scanning shows carbon segregation along dendrite arm boundary of the 0W cast steel [3].

Figure 6.4a and c show the SEM images of the 0W cast steel and the measured positions for EDS analysis. Ten measurements were taken from root to tip along the dendrite arm boundary, and seven measurements were conducted at trigeminal boundaries, as indicated in Figure 6.4a and c respectively. All the measured points were equidistant. The measured concentration of carbon is presented in Figure 6.4b and d. It can be seen that a decreased carbon concentration trend from root to tip along the dendrite arm boundary is found, as shown in Figure 6.4b. The carbon concentration at trigeminal boundary intersection shows higher level than that at other positions, as shown in Figure 6.4d. The average values of measured concentration of carbon at dendrite and grain boundaries are obtained as 8.12 wt.% and 4.60 wt.%, respectively. Carbon concentration at dendrite arm boundary is much higher than that at grain boundaries. Therefore, serious carbon segregation at dendrite arm and grain boundaries has occurred during solidification.

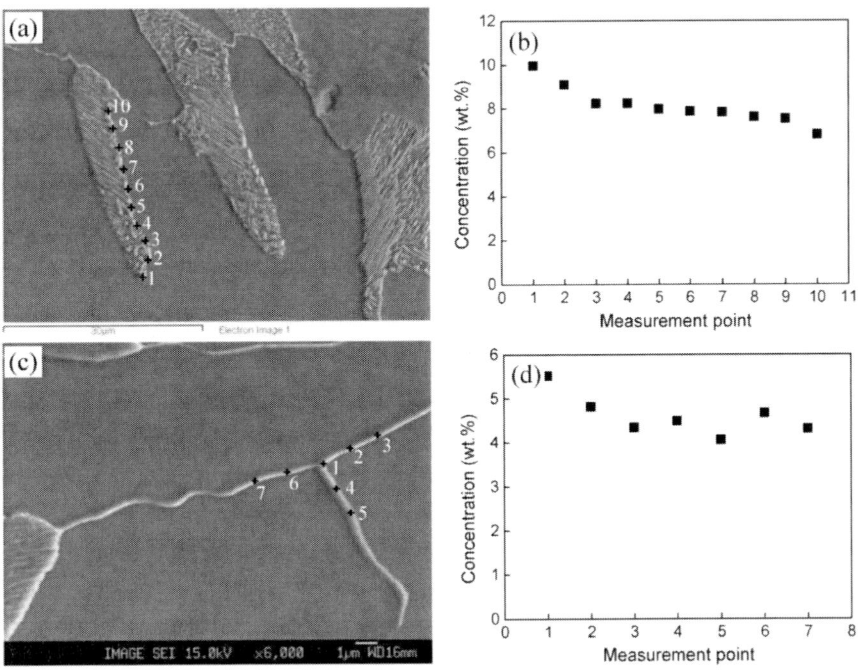

Figure 6.4. Carbon concentration distribution along (a,b) dendrite arm and (c,d) grain boundaries of the 0W cast steel.

The basic cause of segregation is that the liquid rejects solutes during freezing because the solid has less solubility for them as compared to the liquid. If an impure liquid is kept in contact with its solid at the freezing temperature for an appreciable time, equilibrium partitioning of the solutes would be attained between the liquid and the solid. It may then be written as $K=C_S/C_L$, where K is the equilibrium partition ratio, and C_S and C_L are the solute concentrations in the solid and liquid, respectively [2]. In carbon steels, the equilibrium partition ratio of carbon is much lower than that of other elements such as manganese and silicon. Since the solute elements with the lowest K values have the highest tendency to segregate, serious segregation of carbon becomes hard to be avoided during solidification.

Under equilibrium solidification conditions where diffusion in solid and liquid is complete, the compositions of both solid and liquid adjusts during cooling in accordance with the phase diagram so that the final solid has the composition of the original melt, and no carbon segregation occurs. Under real conditions, however, carbon diffusion in the solid is restricted, and the steel after solidification is inhomogeneous in composition. When grains grow parallel to the grains boundaries, grooves form at the contact of solid and liquid during grains growth process.

Carbon tends to be enriched in the grooves, especially at the frontier of solid-liquid interfaces, and then carbon segregation along dendrite arm boundaries will be caused. Under opposite growth condition, highly-enriched carbon exists at the binding-site of grains after solidification, leading to carbon segregation at grain boundaries. The mechanism of carbon segregation at dendrite arm and grain boundaries is schematically illustrated in Figure 6.5a and b, respectively.

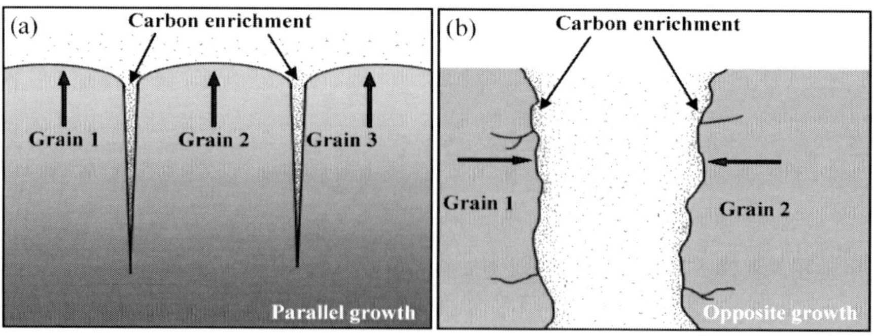

Figure 6.5. Mechanism of carbon segregation at (a) dendrite arm and (b) grain boundaries during solidification.

6.4.2. Heat Treatment Design

As analysed above, the 0W cast steel possesses coarse grains, inhomogeneous microstructure and serious carbon segregation. All these factors work against the improvement of mechanical properties, especially toughness. Experimental evidence also shows that the absorbed Charpy impact energy of the 0W cast steel is only 5.4 J when testing at room temperature. The impact toughness is very poor, and the 0W cast steel cannot be used for practical purposes.

In order to achieve the desired mechanical properties, hot deformation approaches including forging and hot rolling are generally used in low-carbon microalloyed cast steels. Even though hot deformation is an effective method to improve the toughness and strength, the production procedure, energy consumption and then production cost will certainly be increased. If direct processing method, such as heat treatment, after casting is available to improve both toughness and strength to an acceptable level, the problem above can be solved, and the production cost will be undoubtedly reduced. Consequently, demands for producing low-cost, higher strength steel castings with excellent toughness and strength have encouraged many researchers to focus on microalloyed cast steels. Nowadays, microalloyed cast steels have found many applications in the manufacturing of industrial parts such as offshore platform nodes, machinery supports, natural gas compressor housing, ingot moulds and buckets. Since most of these parts have to be heat treated before use, it becomes essential to carry out an investigation on the effects of different heat treatment variables on the microstructure and mechanical properties of microalloyed cast steels with the purpose of enhancing both the toughness and strength to the practical application level.

Prior to heat treatment, two groups of rectangular specimens with size of 12 mm × 12 mm × 60 mm and 12 mm × 12 mm × 110 mm respectively were cut from the cast steel. Annealing treatment was selected to modify the microstructure and improve the toughness due to its advantages described in Section 6.3.1.

Annealing was carried out using an electric furnace in laboratory. Based on the results presented in Table 2.2 of Chapter 2, the A_{c1} and A_{c3} temperatures for the 0W steel are 766°C and 903°C, respectively. Some of the specimens were then initially austenitised at temperatures ranging from 950 to 1200°C in steps of 50°C for 2 h and then cooled in the closed furnace. For comparison, other specimens being heat treated at these temperatures were cooled respectively in air and water after 2 h austenitising. In order to determine the optimal annealing time, some of the specimens were austenitised at 1100°C for 0.5 h and 5 h respectively followed by furnace cooling.

6.4.3. Mechanical Properties after Heat Treatment

Impact properties

Table 6.1 gives the room temperature Charpy impact energies of the cast steel before and after heat treatment [3]. It can be seen that the specimen austenitised at 1100°C for 2 h followed by furnace cooling shows the highest impact energy of 120.3 J, which is about 22 times higher than that of the as-cast one.

Table 6.1. Room temperature impact energies before and after heat treatment [3]

Specimen	Austenitising temperature (°C)	Holding time (h)	Cooling method	Impact energy (J)
HT0	As-cast	-	-	5.4
F950	950	2	Furnace	83.8
A950	950	2	Air	67.2
W950	950	2	Water	17.9
F1000	1000	2	Furnace	89.7
A1000	1000	2	Air	84.9
W1000	1000	2	Water	19.4
F1050	1050	2	Furnace	101.0
A1050	1050	2	Air	77.1
W1050	1050	2	Water	25.2
F1100H0.5	1100	0.5	Furnace	97.9
F1100	1100	2	Furnace	120.3
A1100	1100	2	Air	71.2
W1100	1100	2	Water	27.0
F1100H5	1100	5	Furnace	111.0
F1150	1150	2	Furnace	50.4
A1150	1150	2	Air	44.2
W1150	1150	2	Water	30.7
F1200	1200	2	Furnace	45.6
A1200	1200	2	Air	22.3
W1200	1200	2	Water	26.2

Figure 6.6 shows the dependence of impact energy on heat treatment temperature of the specimens after 2 h austenitisation [3]. It is clear that the impact energy of the specimen cooled in furnace is higher than those cooled in air and water at a fixed temperature. Moreover, the impact energy of the furnace-cooled specimen increases gradually at first, then to a maximum value at the temperature of 1100°C, and finally it drops dramatically when the temperature is higher than 1100°C.

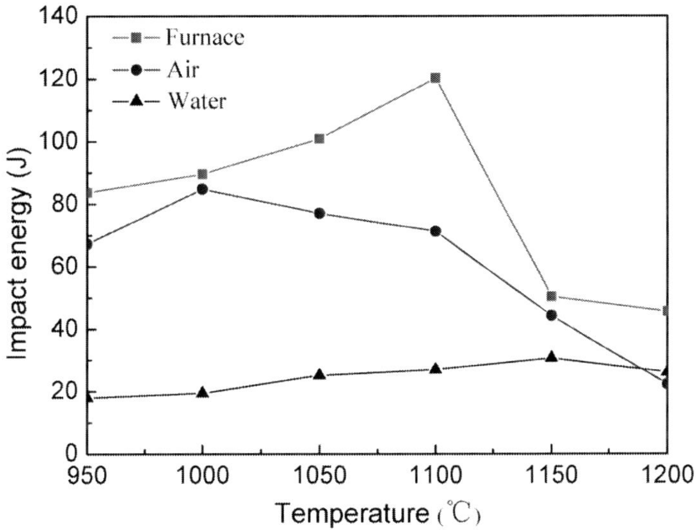

Figure 6.6. Plots of impact energy versus heat treatment temperature at various cooling conditions after 2 h austenitisation [3].

Tensile properties

Room temperature tensile tests were performed on the furnace-cooled specimens. The typical tensile curves are shown in Figure 6.7 [3]. The tensile properties obtained from Figure 6.7 are presented in Table 6.2 [3]. The results in Figure 6.7 indicate that the tensile behaviour of the tested specimens is similar. Based on the results shown in Table 6.2, the specimen austenitised at 950°C for 2 h and cooled in furnace (F950) shows the highest YS (378.5 MPa) and UTS (541.6 MPa) with elongation of 29.4%, while the specimen austenitised at 1100°C for 2 h and cooled in furnace (F1100) shows the highest value of elongation (34.3%) with YS and UTS of 344.9 MPa and 511.3 MPa, respectively. However, taking the impact properties into consideration, the specimen F1100 shows higher impact resistance than that of the specimen

F950. To evaluate the mechanical properties, other measurements, such as the microstructure and fracture surface, are therefore necessary to be carried out on the specimens.

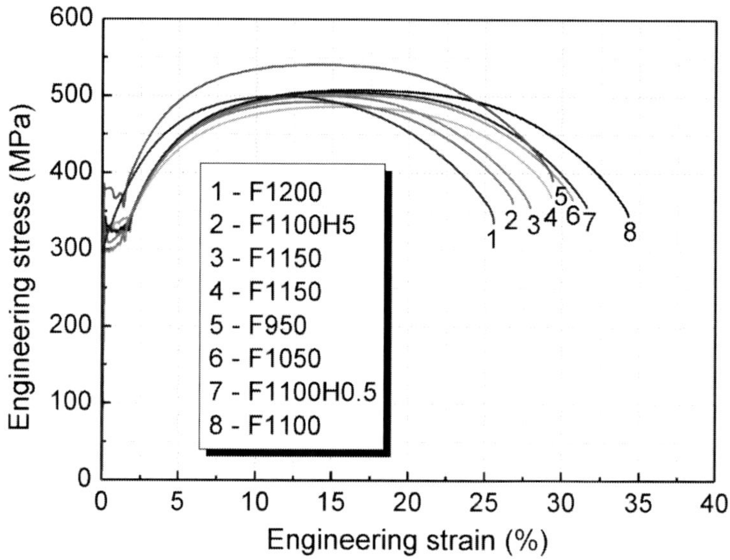

Figure 6.7. Typical tensile curves of the furnace-cooled specimens [3].

Table 6.2. Tensile properties of the furnace-cooled specimens [3]

Specimen	YS (MPa)	UTS (MPa)	Engineering strain (%)
F950	378.5	541.6	29.4
F1000	328.8	486.2	29.4
F1050	341.1	501.9	30.8
F1100H0.5	335.4	505.1	31.6
F1100	344.9	511.3	34.3
F1100H5	322.1	492.5	26.9
F1150	299.2	500.1	27.9
F1200	310.2	499.8	25.6

6.4.4. Microstructural Evolution after Heat Treatment

Microstructural changes

Figure 6.8 shows the OM structures of the furnace-cooled specimens after 2 h austenitisation [3]. It can be seen that all the microstructures consist of equiaxed ferrite and pearlite. Austenitising temperature has a significant effect on the distribution of pearlite. When the temperature is low, the pearlite shows a ring-shaped distribution feature, as shown in Figure 6.8a.

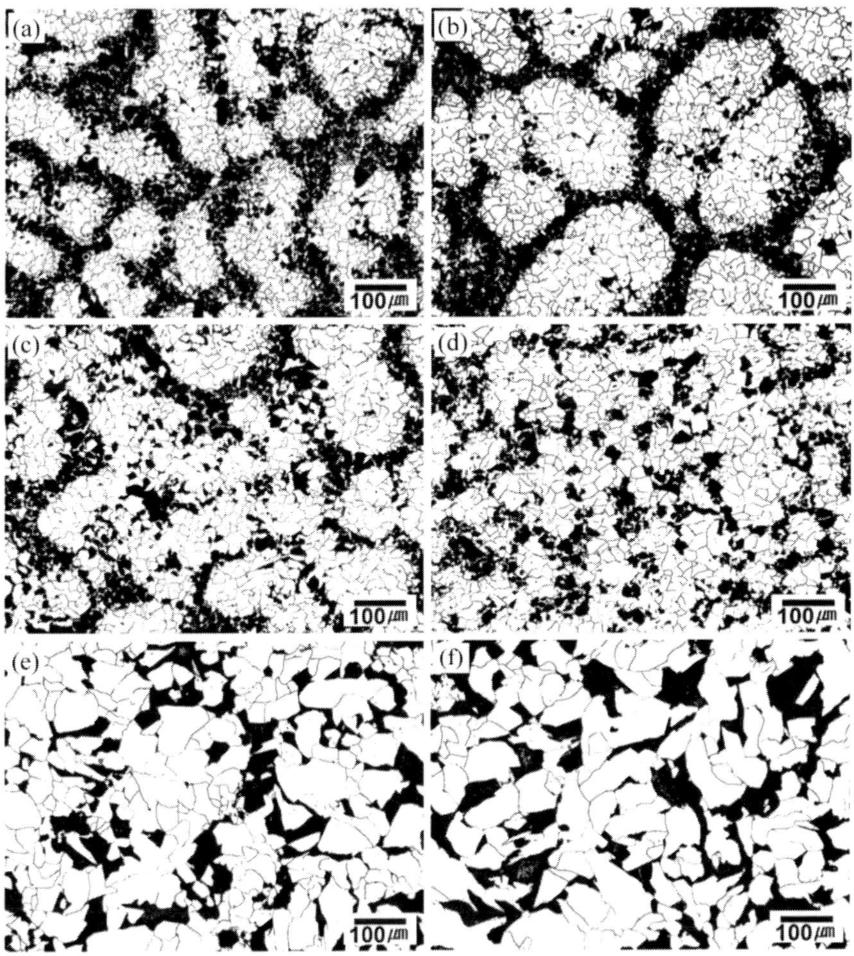

Figure 6.8. OM structures of the furnace-cooled specimens after 2 h austenitisation: (a) F950, (b) F1000, (c) F1050, (d) F1100, (e) F1150, and (f) F1200 [3].

With an increase of temperature, the ring-shaped pearlite cluster tends to be broken up into separate ones, and the pearlite distribution becomes more and more uniform, as shown in Figure 6.8b, c and d. Although homogenised microstructures can be obtained when the specimens are heat treated at temperatures of 1150°C and 1200°C (Figure 6.8e and f), the negative effect on grain growth is clear as compared to the specimens heat treated at lower temperatures.

Figure 6.9 shows the OM structures of the furnace-cooled specimens after 0.5 h and 5 h austenitisation at 1100°C [3]. The effect of holding time on the pearlite distribution can be clearly seen in Figures 6.8d and 6.9. Obviously, short holding time is not effective in homogenising the microstructure, as shown in Figure 6.9a. If the holding time is long (5 h) however, some of the grains grow abnormally, as shown in Figure 6.9b.

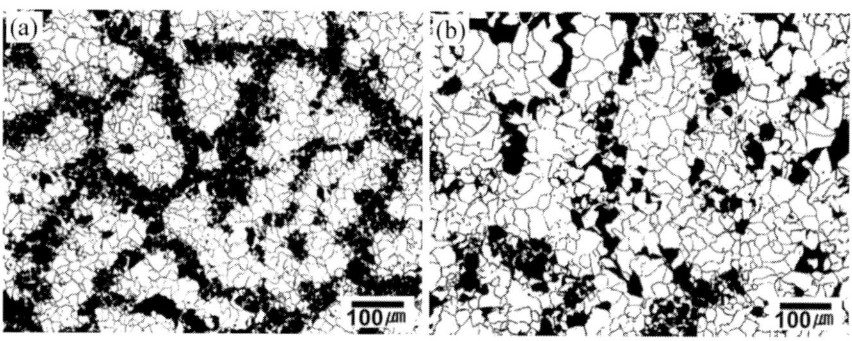

Figure 6.9. OM structures of the furnace-cooled specimens after (a) 0.5 h and (b) 5 h austenitisation at 1100°C [3].

The OM structures of the specimens that austenitised at 1100°C for 2 h followed by air cooling and water quenching are shown in Figure 6.10a and b, respectively [3].

As shown in Figure 6.10a, the microstructure of the air-cooled specimen is mainly composed of polygonal ferrite and pearlite, accompanied with small quantity of Widmanstätten ferrite. Widmanstätten ferrite has a coarse, elongated morphology, readily resolved in the light microscope. Widmanstätten ferrite forms at faster cooling rates than polygonal ferrite and in temperature ranges just below those at which equiaxed ferrite forms. The microstructure of the water quenched specimen consists of full martensite, as shown in Figure 6.10b.

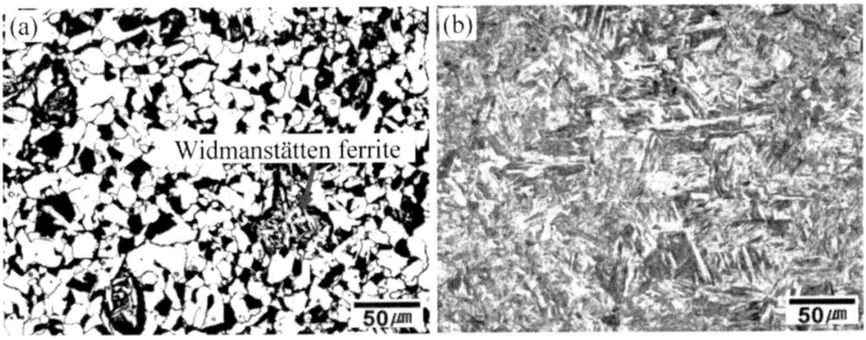

Figure 6.10. OM structures of the (a) air-cooled and (b) water-quenched specimens after 2 h austenitisation at 1100°C [3].

Figure 6.11a and b show the variations of ferrite grain size with temperature and holding time, respectively [3]. As demonstrated in Figure 6.11a, the ferrite grain size increases slightly with temperature in the temperature range of 950 to 1100°C, then suddenly increase in ferrite grain size when the temperature is higher than 1100°C. According to the result of Figure 6.11b, the effect of holding time on ferrite grain size is insignificant. With regard to Figures 6.8, 6.9 and 6.11, the specimen F1100 shows the best combination of uniformly distributed microstructure and fine ferrite grains.

Discussion on the effect of heat treatment on microstructural evolution

Considerable diffusion of carbon is necessary for austenite formation to balance the carbon in various phases of the reaction. Cementite is the source of carbon for the austenite nucleation, and therefore the reaction begins at carbon rich ferrite/cementite interfaces. After nucleation, the further growth is still dependent on carbon diffusion through the surrounding austenite to austenite/ferrite interfaces. So, the distribution and diffusion of carbon have an important effect on the austenite formation and growth.

Based on the results shown in Figures 6.3 and 6.4, severe carbon segregation has been caused during solidification, leading to non-uniform distribution of carbon in the as-cast specimen. When the as-cast specimen is heated to the austenitising temperature, austenite formes accompanied with carbon diffusion in the matrix. Because the diffusion is strongly influenced by temperature, limited carbon diffusion occurs if the austenitising temperature is low, and non-uniformly distributed austenite is then considered to be formed. Figure 6.12 shows the prior austenite grains of the water-quenched specimen after holding at 950°C for 2 h [3]. It can be seen that the distribution of prior austenite grains is non-uniform. Coarse grains, around which lots of fine grains are observed, distribute discontinuously in the matrix.

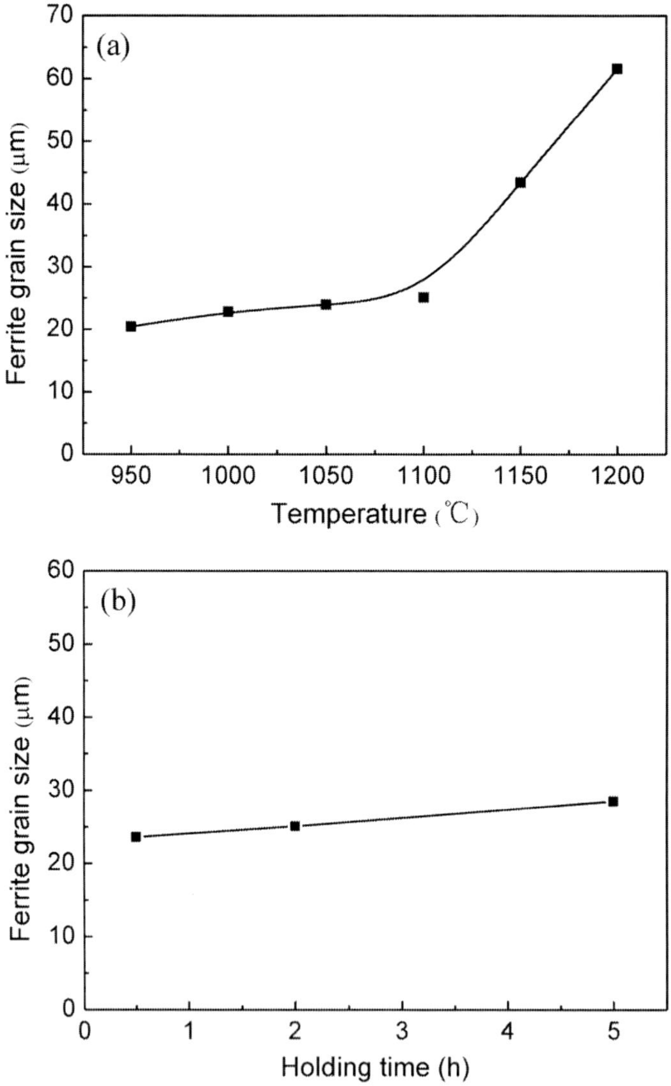

Figure 6.11. Effect of temperature and holding time on ferrite grain size of the heat treated specimens after furnace cooling with (a) holding time of 2 h and (b) austenitising temperature of 1100℃ [3].

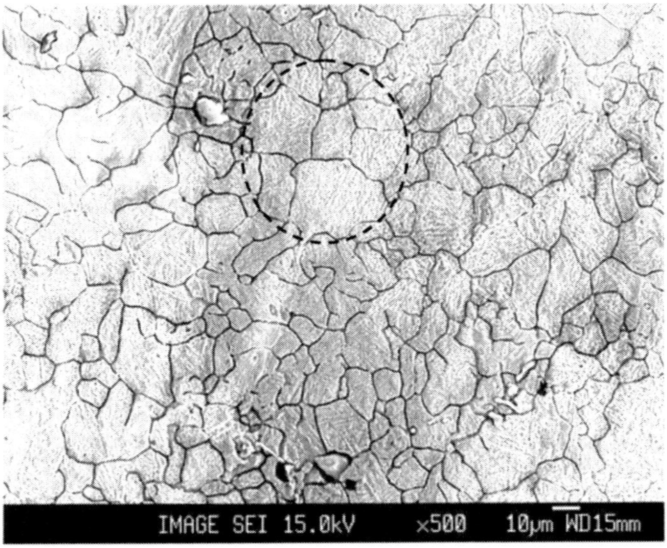

Figure 6.12. SEM image of the specimen W950 shows non-uniform distribution of prior austenite grains [3].

Pearlite is composed of alternate layers of cementite and ferrite. For slowly cooled microalloyed steel, the pro-eutectoid ferrite reaction precedes the pearlite reaction. Pearlite forms from austenite by a nucleation and growth process. It is generally accepted that pearlite nucleates preferentially at austenite grain boundaries [4]. For an austenite structure with uniform composition and grains, the free energy available for the formation of nuclei is the same in every volume element. Accordingly, the nucleation probability of pearlite is same at different austenite grain boundaries, and uniform distributed ferrite-pearlite microstructure will be obtained. In the case of non-uniform distribution of austenite grains, fine austenite grains supply greater available number of nucleation sites and higher activation energy for nucleation compared to that of coarse grains. For reducing free energy, the boundaries of fine austenite grains will be the preferred sites for pearlite nucleation. Moreover, the austenite-to-pearlite transformation rate increases with a decrease of austenite grain size due to an increase in nucleation rate of pearlite. In other words, the regions with fine grains preferentially nucleate when austenite-to-pearlite phase transformation occurs. Finally, as schematically represented in Figure 6.12 with dark circle (fine austenite grains run around the coarse ones), ring-shaped distribution characteristic of pearlite will be displayed after austenite-to-pearlite phase transformation (Figure 6.8a). With the increase of austenitising temperature and holding time, austenite distribution becomes more and more uniform, and subsequently, uniformly distributed pearlite will be available. The higher the austenitising temperature and the longer the holding time, the more uniform the pearlite distribution will be, as shown in Figures 6.8 and 6.9.

Many characteristics in steel, such as mechanical properties and the microstructures of austenite transformation products, depend on the austenite grain size. In microalloyed steel, the austenite grain size has a critical effect on the formation of fine-grained ferritic microstructure with high toughness and strength.

As discussed in Chapter 4, the austenite grain size of the 0W steel slightly increases as the temperature increases from 950 to 1100°C. When the temperature is higher than 1100°C, sharp increase in austenite grain size is observed (see Figure 4.3 of Chapter 4). Moreover, holding time shows little influence on the austenite grain size (see Figure 4.5 of Chapter 4). According to Figures 4.3, 4.5 and 6.11, it is clear that ferrite grain size shows the same variable tendency with that of austenite grain size.

6.4.5. Fractography

Selected fracture surfaces in the shear lip areas of some impact test specimens are demonstrated in Figure 6.13 [3]. As can be seen in Figure 6.13a, the fracture surface of the specimen HT0 is composed of large cleavage facets and shear cracks, which indicates the brittle nature of fracture. The fracture mode has been remarkably changed after heat treatment. The fracture surface of the specimen F950 is a mixture of large dimples and small cleavage facets, as shown in Figure 6.13b. Small, deep and uniformly distributed dimples are observed in the fracture surface image of specimen F1100, which shows a characteristic of ductile fracture (Figure 6.13c). However, as shown in Figure 6.13d, the fracture mode of the specimen that austenitised at 1150°C for 2 h and cooled in furnace (F1150) becomes cleavage-type again. According to these results, a critical temperature is necessary for obtaining ductile mode fracture. Moreover, neither shorter nor longer holding time are favourable to generate small, deep and uniformly distributed dimples after impact tests (Figure 6.13e and f).

According to the above results, when the temperature is lower than 1100°C, the uniform distribution of pearlite after heat treatment significantly affects the impact energy. Slight increase in ferrite grain size with temperature shows an inconspicuous effect on the impact energy. Sharp decrease of impact energy is caused by the rapid growth of ferrite grains as the temperature is elevated to 1150°C.

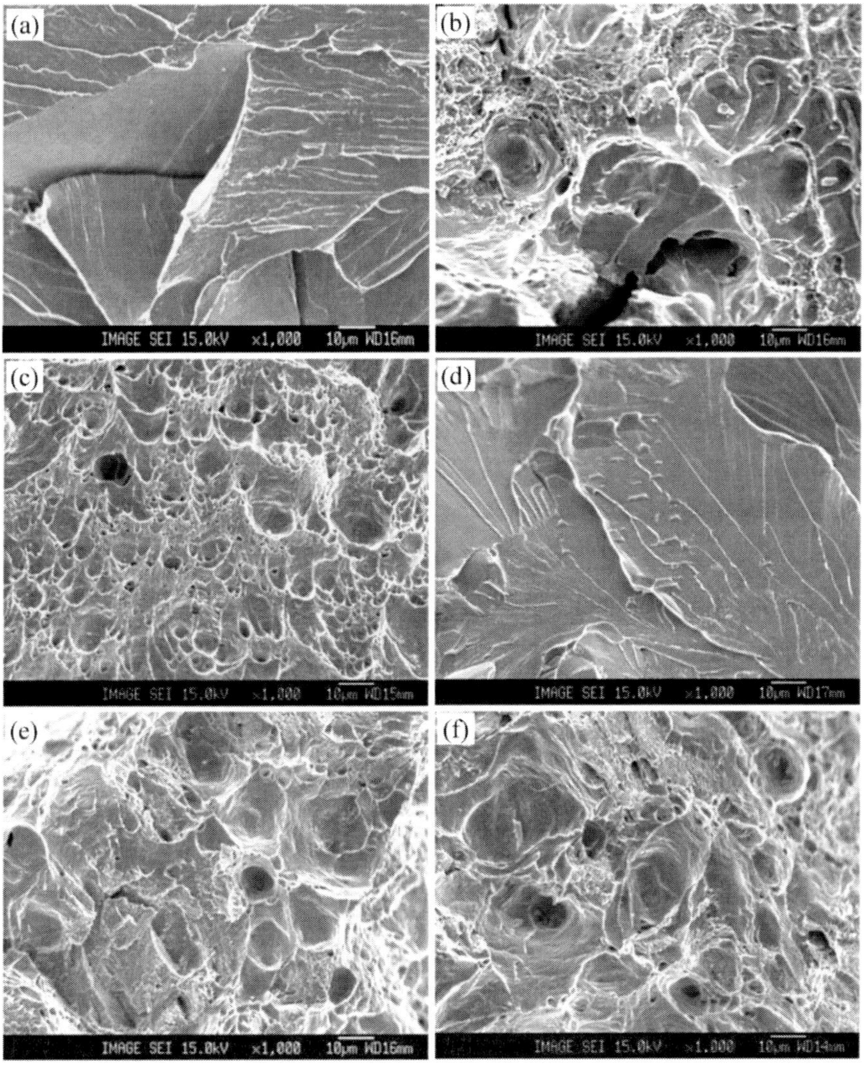

Figure 6.13. Selected fracture surfaces in the shear lip areas of some impact test specimens: (a) HT0, (b) F950, (c) F1100, (d) F1150, (e) F1100H0.5, and (f) F1100H5 [3].

Intact and small-sized ring of pearlite distributed at 950°C is of advantage to increase the tensile strength. In the temperature range of 950 to 1100°C, the more uniform the pearlite distribution, the higher the elongation is. Moreover, coarsening of ferrite grain size has a negative effect on the enhancement of engineering strain. Taking into account the impact energy, tensile strength, elongation, microstructure and fracture morphology, the specimen F1100 exhibits the best combination of excellent mechanical properties and optimal microstructure. Compared with the European Standard [5], the mechanical properties of F1100 show a good fit to the requirement of S275 and S355 steels. Even though for S420 steels, the required mechanical properties can also be satisfied in the large nominal thickness range.

6.5. Heat Treatment of Microalloyed Forging Steels

6.5.1. Why Heat Treatment is Essential?

Many applications for structures, machine and vehicle parts require high strength and excellent toughness. Traditionally, steels for these applications are produced by forging, and the required properties are obtained by quenching and tempering treatment. Interest has increased in the use of microalloyed forging steels in place of quenched-and-tempered steels to realise cost savings. Microalloyed forging steels have been developed over the past several years and are increasingly applied in a variety of large-scale structures, engine, transmission, suspension and driveline components such as the flange, gears, springs, crankshafts, connecting rods, axles and shaft.

Microalloyed forging steels employ small amount of microalloying elements to improve microstructure and mechanical properties, or to enable alternative processing routes. The required strength level is mainly provided by grain refinement and/or precipitation strengthening from microalloying elements. Undissolved and precipitated carbides, nitrides or carbonitrides restrict the movement of either dislocations or interfaces and prevent grain coarsening during heating and processing, resulting in improved strength. Compared with conventional quenched-and-tempered steels, microalloyed forging steels exhibit not only simplified operation procedures, but also increased microstructure homogeneity, less stress concentration and improved dimensional stability.

Decreased toughness, however, is usually accompanied. The results of room temperature Charpy impact tests shown in Table 5.1 of Chapter 5 also indicate that the absorbed impact energies of all the microalloyed forging steels with and without tungsten additions are very poor. As a result, the application of these steels will undoubtedly be limited based on toughness consideration at the required strength level. Therefore, in order to acquire excellent toughness, appropriate heat treatment is essential and of significance to be performed on these microalloyed steels after hot forging.

The air-cooled microstructure of microalloyed forging steels is generally composed of ferrite and pearlite. The content, size and distribution of ferrite and pearlite within the microstructure play an important role in the final mechanical properties. As a strong ferrite former, tungsten can restrain pearlite and ferrite transformations, and is effective for precipitate refining and solid solution strengthening.

It implies that complicated microstructural evolution in the tungsten-containing microalloyed steels will be involved during heat treatment, and the resultant microstructure may exhibit a different feature as compared to the tungsten-free steels after heat treatment. Since the variations in mechanical properties of steel after heat treatment are caused by the phase transformations and structural changes that occur during the heat treatment, the optimal heat treatment conditions should be different for the steels with different tungsten additions due to the functions of tungsten alloying in them. Thus, to find optimal heat treatment processes for the microalloyed forging steels with different tungsten additions with the purpose of achieving the best combination of required strength levels and acceptable toughness, it is essential to conduct a systematic heat treatment investigation on them to understand well the relationship between compositions, microstructure and mechanical properties after heat treatment.

6.5.2. Heat Treatment Design

Two groups of rectangular specimens with sizes of 11 mm × 11 mm × 60 mm and 11 mm × 11 mm × 110 mm respectively were cut from the forged pieces. Specimens were heat treated at 840, 860, 880, 900, 920 and 950°C for 1 h using an electric furnace in laboratory followed by air and furnace cooling based on the measured critical temperatures of A_{c1} and A_{c3} that presented in Table 2.2 of Chapter 2. The heat treatment schedule is schematically shown in Figure 6.14.

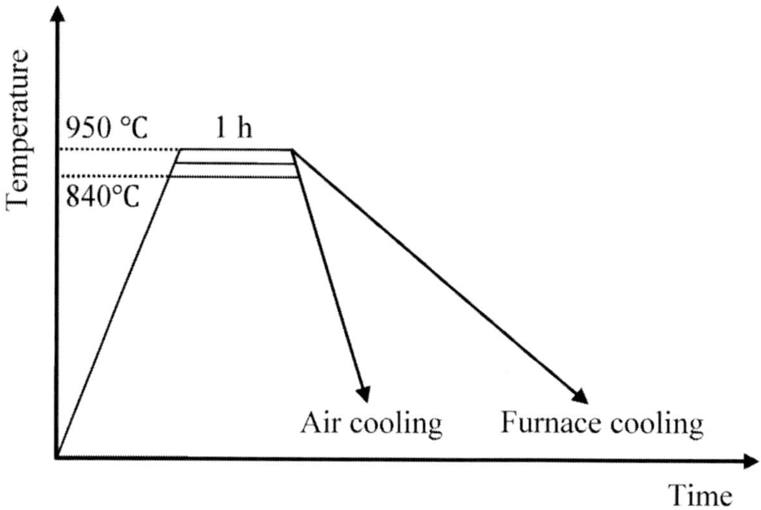

Figure 6.14. Schematic illustration of the heat treatment schedule for the microalloyed forging steels.

6.5.3 Effect of Tungsten on the Mechanical Properties after Heat Treatment

Impact properties

Figure 6.15a and b show the dependence of impact energies on heat treatment temperatures under air and furnace cooling conditions, respectively [6]. For comparison, the impact energies of the as-forged specimens are also presented in Figure 6.15. It can be seen that the toughness of the as-forged specimens has been significantly enhanced after heat treatment. The absorbed impact energy of the heat-treated 0W specimen suffered from air cooling exhibits higher value than that suffered from furnace cooling, while the heat-treated tungsten-containing specimens subjected to furnace cooling show higher toughness level than that subjected to air cooling. The highest impact energies of the heat-treated 0W, 0.5W, 1W and 2W specimens suffered from air cooling are 208.3, 66.0, 51.9 and 23.4 J, and the corresponding heat treatment temperatures are 920, 900, 880 and 860°C, respectively. Differently, all the furnace-cooled specimens exhibit their highest toughness levels at 880°C, and the maximum impact energies of the heat-treated 0W, 0.5W, 1W and 2W specimens are 149.6, 120.3, 82.2 and 47.2 J, respectively.

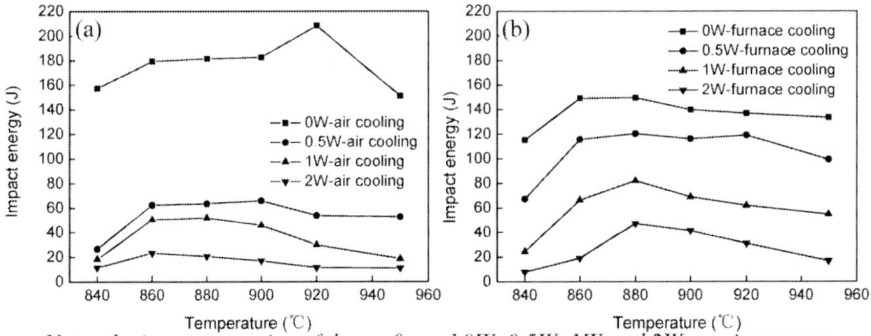

Note: the impact energies of the as-forged 0W, 0.5W, 1W and 2W specimens are 19.3, 14.5, 10.6 and 9.8 J, respectively.

Figure 6.15. Dependence of impact energies on temperatures under (a) air and (b) furnace cooling conditions [6].

Tensile properties

The heat-treated specimens with the highest toughness levels under both air and furnace cooling conditions were employed to do the tensile tests. The typical tensile curves are presented in Figure 6.16 [6]. The tensile properties determined from these curves are listed in Table 6.3 [6].

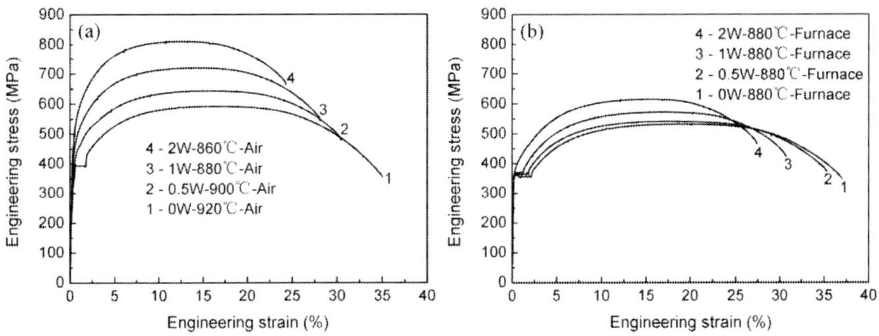

Figure 6.16. Typical tensile curves of the specimens after heat treatment followed by (a) air and (b) furnace cooling [6].

The result shown in Figure 6.16a indicates that the addition of tungsten leads to continuous yielding when the heat-treated specimens suffered from air cooling. Under furnace cooling condition, the tensile behaviour of the 2W specimen with continuous yielding is different from that of other three specimens in which apparent yield points can be observed, as shown in Figure 6.16b. Based on Table 6.3, the heat-treated specimens subjected to air cooling exhibit higher strength levels than that subjected to furnace cooling at the expense of strain. According to Table 5.2 of Chapter 5 and Table 6.3, it is apparent that both the tensile strength and strain of the as-forged specimens have been significantly increased after heat treatment followed by air cooling. Under furnace cooling condition, however, decreased tensile strength of the as-forged specimens is caused, even though the strain has been greatly increased.

Table 6.3. Tensile properties of the specimens after heat treatment [6]

Specimen	Heat treatment temperature (°C)	Cooling method	YS (MPa)	UTS (MPa)	Engineering Strain (%)
0W	920	Air	390.1	593.2	35.0
0.5W	900	Air	431.1	655.7	30.5
1W	880	Air	476.9	724.7	28.1
2W	860	Air	532.6	811.6	24.3
0W	880	Furnace	356.3	533.1	37.0
0.5W	880	Furnace	364.7	542.6	35.3
1W	880	Furnace	382.2	572.7	30.7
2W	880	Furnace	409.5	615.5	27.4

6.5.4. Effect of Tungsten on the Microstructure after Heat Treatment

The microstructures of the air-cooled specimens according to Table 6.3 are presented in Figure 6.17 [6]. As can be seen, the microstructure of the air-cooled 0W specimen consists of equiaxed ferrite and pearlite, and the grain size distribution is very uniform (Figure 6.17a).

According to Figure 4.9a of Chapter 4 and Figure 6.17a, it is apparent that heat treatment has significantly refined the microstructure of the as-forged 0W specimen. In contrast to that in the 0W specimen, small amount of bainite appears in the microstructure of the 0.5W specimen (Figure 6.17b). With an increase of the tungsten content, the amount of bainite increases, as shown in Figure 6.17c and d. In addition, the size of bainite packets has been greatly refined after heat treatment.

Figure 6.18 illustrates the microstructures of the furnace-cooled 1W and 2W specimens according to Table 6.3 [6]. It can be seen from Figure 6.18a that the microstructure of the furnace-cooled 1W specimen is mixed by equiaxed ferrite and pearlite. Differently, small amount of bainite appears in the microstructure when the tungsten content is increased to 2%, as shown in Figure 6.18b. Moreover, furnace cooling induces coarser grains as compared to that under air cooling condition at the same heat treatment temperature, as presented in Figures 6.17c and 6.18a.

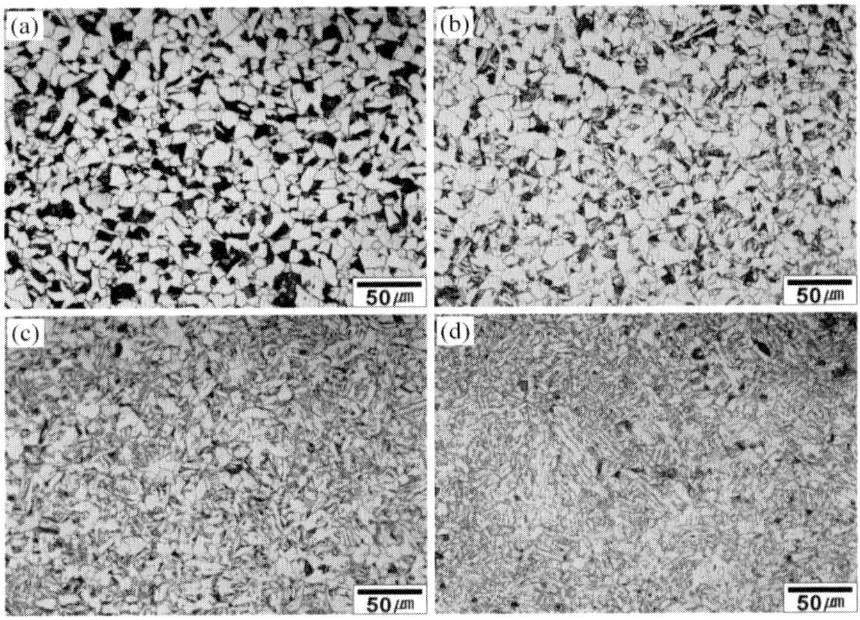

Figure 6.17. Microstructures of the air-cooled specimens: (a) 0W + 920°C, (b) 0.5W + 900°C, (c) 1W + 880°C and (d) 2W + 860°C [6].

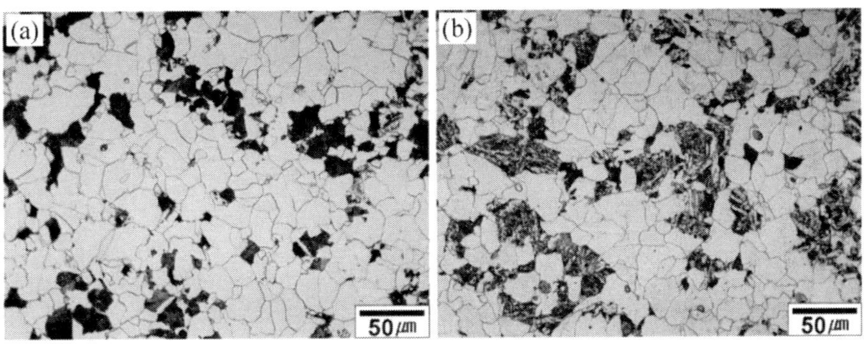

Figure 6.18. Microstructures of the furnace-cooled specimens: (a) 1W + 880°C, and (b) 2W + 880°C [6].

6.5.5. Fractography

Selected fracture surfaces in the shear lip areas of some impact specimens are demonstrated in Figure 6.19 [6]. It can be seen from Figure 6.19a that the as-forged 0W specimen exhibits a brittle cleavage fracture. The fracture surface of the heat-treated 0W specimen suffered from air cooling is composed of small, deep and uniformly distributed dimples, which indicates the ductile nature of fracture, as shown in Figure 6.19c.

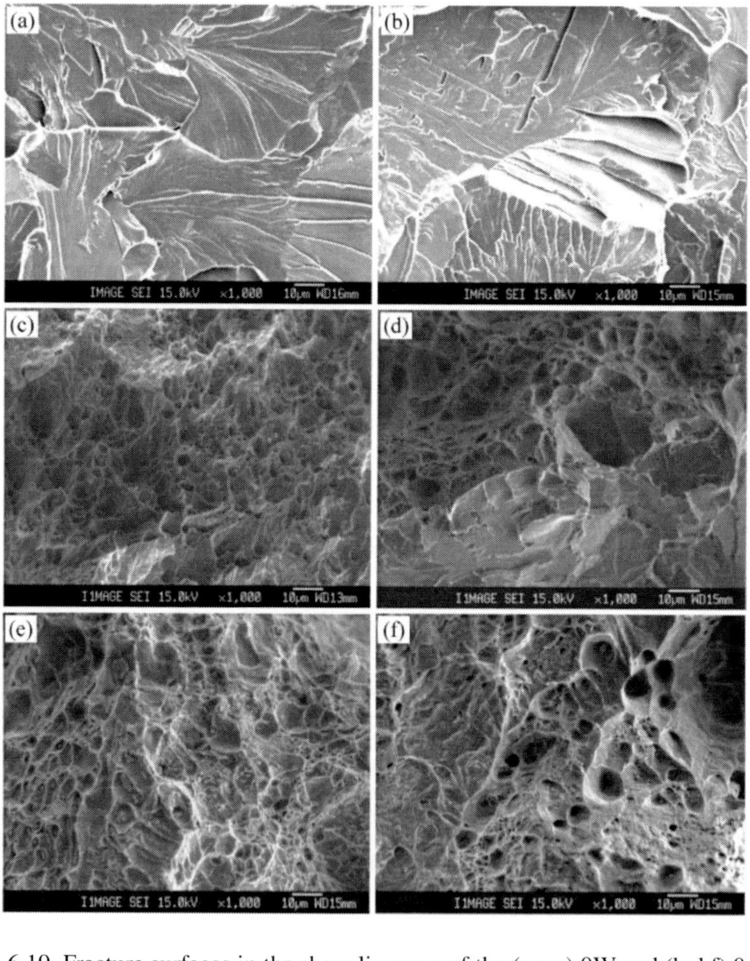

Figure 6.19. Fracture surfaces in the shear lip areas of the (a,c,e) 0W and (b,d,f) 0.5W specimens after Charpy impact tests: (a,b) as-forged, (c) 920°C + air cooling, (d) 900°C + air cooling, and (e,f) 880°C + furnace cooling [6].

Based on Figure 6.19c and e, furnace cooling is found to lead to larger dimples as compared to that suffered from air cooling. The fracture surface of the as-forged 0.5W specimen is a mixture of large cleavage facets and shear cracks, which also shows a characteristic of brittle fracture, as shown in Figure 6.19b. After heat treatment followed by air cooling, the fracture surface of the 0.5W specimen becomes a mixture of cleavage facets and small dimples, as shown in Figure 6.19d. However, as shown in Figure 6.19f, a ductile mode fracture is obtained when the heat-treated 0.5W specimen is suffered from furnace cooling.

Figure 6.20 compares the tensile fracture morphologies of the 0W, 0.5W and 2W specimens before and after heat treatments [6]. It can be seen that all the fracture surfaces are composed of dimples, which indicate a characteristic of ductile fracture. The size of dimples exhibits a decreasing trend with the increase of tungsten content (Figure 6.20a, c and e). Moreover, as shown in Figure 6.20b, d and f, heat treatment has a beneficial effect on reducing the size of dimples.

6.5.6. Discussion on the Mechanical Behaviour after Tungsten Addition

The yield strength of a polycrystalline solid increases in a linear manner with $d^{-1/2}$ (d is the grain size), and ductile to brittle transition temperature decreases in proportion to $d^{-1/2}$ [7]. Thus, grain refinement increases both strength and toughness. After heat treatment followed by air cooling, both the strength and toughness of the as-forged 0W specimen are significantly enhanced.

According to Figure 4.9a of Chapter 4 and Figure 6.17a, these enhanced strength and toughness levels are closely related to the remarkable ferrite grain refinement. Since the A_{c3} temperature of the 0W specimen is 903°C, heat treatment at temperatures lower than 903°C induces incomplete austenitisation, leading to non-uniform distribution of ferrite. Coarser ferrite grains will be caused if the austenitising temperature is higher than 920°C due to the increase of austenite grain size with temperature. These will consequently reduce the toughness. Under furnace cooling condition, the reduction in impact energy for the 0W specimen should be attributed to the coarsening of ferrite grains during the very slow cooling process.

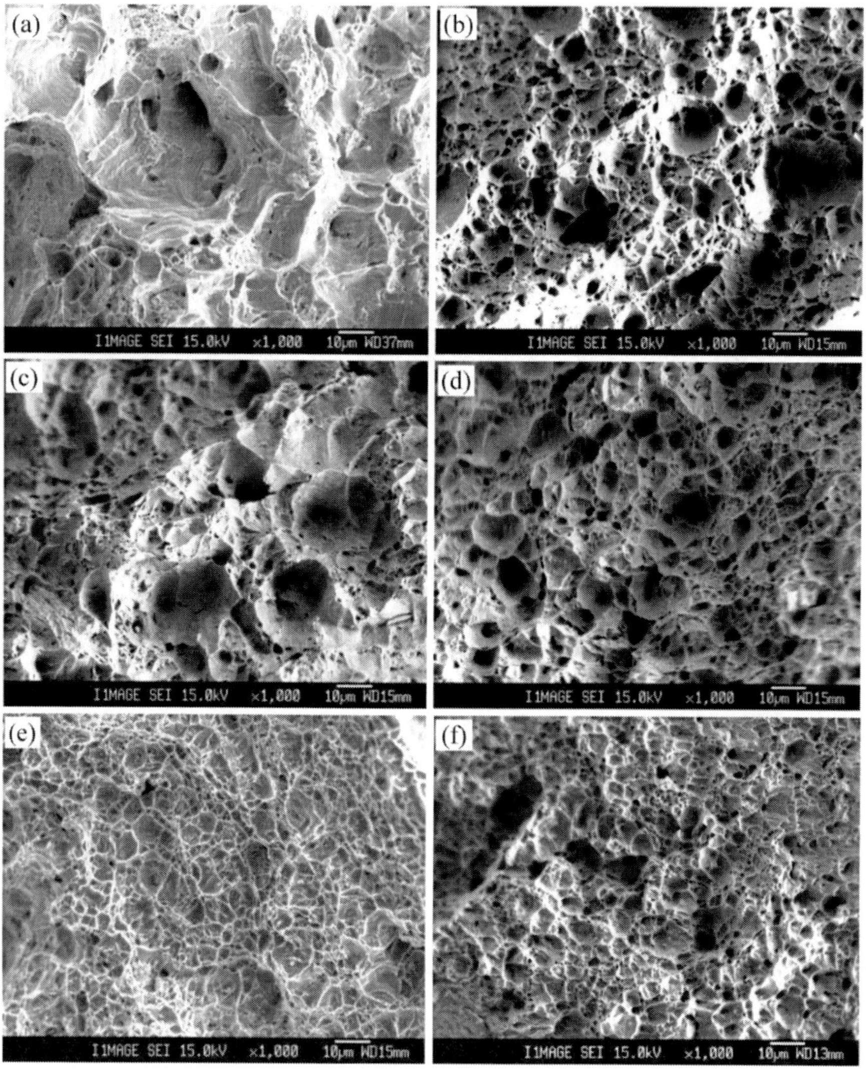

Figure 6.20. Comparison of tensile fracture morphologies of (a,b) 0W, (c,d) 0.5W and (e,f) 2W specimens before and after heat treatment: (a,c,e) as-forged, (b) 920°C + air cooling, (d) 900°C + air cooling, and (f) 860°C + air cooling [6].

The hardenability increases after tungsten addition, promoting the formation of bainite [8]. Because bainite does not facilitate the enhancement of toughness but strength compared with ferrite, the formation of bainite under air cooling condition is responsible for the lower toughness and higher strength than that suffered from furnace cooling. In contrast to the optimal temperatures corresponding to the highest toughness values under air cooling condition, higher temperature induces larger bainite packet, and lower temperature leads to the formation of non-uniformly distributed microstructure, as shown in Figure 6.21 [6]. During cleavage failure, the cracks propagate undeviated across individual packets of bainite [9], and the size of cleavage facets obtained by brittle fracture correlates well with the size of the bainite [10]. These microstructural morphologies formed at either higher or lower temperature is, therefore, unfavourable to enhance the toughness. Therefore, there is an optimal temperature for obtaining the excellent mechanical properties, and the microstructural characteristic plays important roles in the enhancement of both impact toughness and tensile strength.

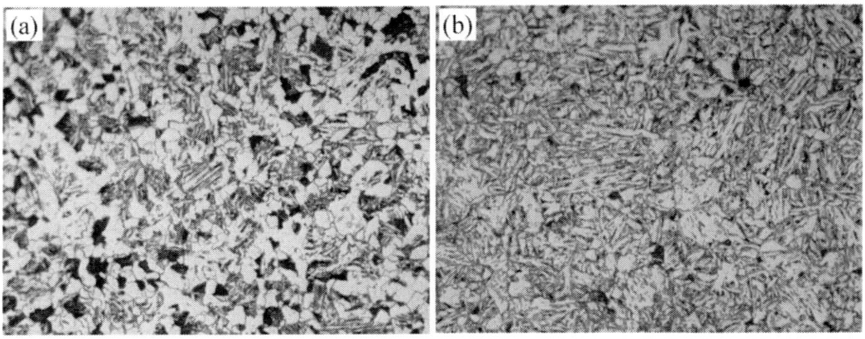

Figure 6.21. Microstructures of the heat-treated 1W steel after air cooling: (a) 840°C, and (b) 950°C [6].

Tensile curves of the air-cooled 0W and the furnace-cooled 0W, 0.5W and 1W specimens after heat treatment exhibit discontinuous yielding behaviour. According to the results of microstructural observation, all these specimens present the same microstructural characteristic mixed by ferrite and pearlite. Differently, bainite exists in all the microstructures of other specimens who have continuous yielding behaviour. The factors which govern discontinuous yielding in steels are [11,12]: (I) low mobile dislocation density of the order of $10^2 \sim 10^4$ cm^{-2} prior to deformation, (II) a rapid rise in the number of dislocations that occur during deformation, and (III) a strong dependence of the dislocation rate on the applied stress. To achieve discontinuous yielding, pinning of dislocations during deformation is essential. Cottrell [13] established the basis of the pinning of dislocations by interstitial atoms in ferrite. According to this theory, the pinning of dislocations by segregated interstitial atoms in ferrite is so strong effectively that all dislocations are immobilised. For plastic deformation to proceed, new dislocations must be generated, and a sufficient stress is applied. As a result, discontinuous yielding in the ferrite-pearlite specimens will appear. In the bainite-containing specimens, the majority of dislocations produced during bainite transformation are mobile in nature [14], even though a considerable number of dislocations can be generated by the transformation of austenite to bainite. Therefore, the observed continuous yielding behaviour of the bainite-containing specimens is attributed to the possibility that a high density of mobile dislocations is produced due to the formation of bainite in the microstructures.

6.6. Practical Application of Heat Treatment of Heavy Sectional Microalloyed Steel

In this section, a heat treatment technology is introduced and applied to a real heavy sectional microalloyed steel produced by ring-forging with the purpose of acquire the uniformly distributed chemical compositions, microstructure and mechanical properties.

Heavy sectional steel production and heat treatment design

The designed chemical compositions of the microalloyed steel are presented in Table 6.4. Casting ingot was heated to 1230°C for a period of time and then ring-forged (or called ring-rolled) to be a heavy sectional wind tower flange with dimensions of 3.7 m in inner diameter, 0.15 m in thickness and 0.19 m in height followed by air cooling. The holding time was determined by 1 h per 50 mm thickness. The final forging temperature was controlled to be higher than 950°C. The forging operation of the heavy sectional flange was carried out in a steel forging company.

Table 6.4. Designed chemical compositions of the microalloyed steel (wt.%)

C	Mn	Cr	Ni	Cu	Si	Al	V	Nb	Co	Mo	P	S
0.17	1.3	0.15	0.25	0.1	0.2	0.03	0.02	0.04	0.01	0.05	0.01	0.005

The flange was reheated to 880°C at a rate of 100°C/h, and then held for 4 h followed by air cooling. Heat treatment was carried out using a large-scale industrial furnace. For analysing the microstructure and mechanical properties after heat treatment, samples were prepared from the surface and central zones along longitudinal (LS, LC), thickness (TS, TC) and radius (RS, RC) directions of the flange, as shown in Figure 6.22.

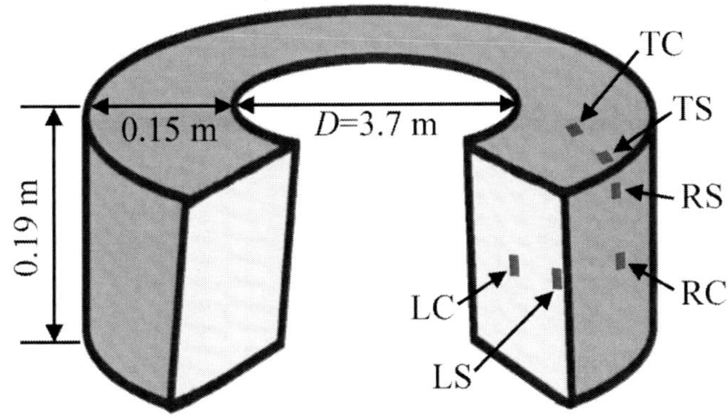

Figure 6.22. Dimension of the wind tower flange and the sampling locations.

Uniformity of real chemical compositions

Table 6.5 shows the real chemical compositions of the forged steel after heat treatment. It can be seen that the locations of LS and LC have almost the same chemical compositions, indicating uniformly distributed chemical compositions of the heat treated steel have been obtained.

Table 6.5. Real chemical compositions of the microalloyed steel (wt.%)

Location	C	Mn	Cr	Ni	Cu	Si	Al	V	Nb	Co	Mo	P	S
LS	0.158	1.34	0.154	0.256	0.101	0.232	0.029	0.023	0.037	0.008	0.049	0.013	0.005
LC	0.160	1.33	0.154	0.258	0.101	0.233	0.029	0.023	0.036	0.008	0.049	0.013	0.005

Microstructural distribution

Figure 6.23 shows the microstructures of the heat treated steel observed at locations prepared along longitudinal, thickness and radius directions of the flange. It can be seen that all the microstructures consist of ferrite and pearlite, and the distributions of ferrite and pearlite are very uniform.

Along the longitudinal direction, the microstructures in both the surface and central zones show similar grain size, as shown in Figure 6.23a and b. For the specimens prepared along the thickness and radius directions, finer grains are observed in the surface zones (Figure 6.23c and e) relative to that in the central zones (Figure 6.23d and f).

In the surface zones, the grain sizes in specimens RS and TS are smaller than that in specimen LS, as shown in Figure 6.23a, c and e. In the central zones, the sizes of some grains in specimens RC and TC slightly increase relative to that in specimen LC, as shown in Figure 6.23b, d and f.

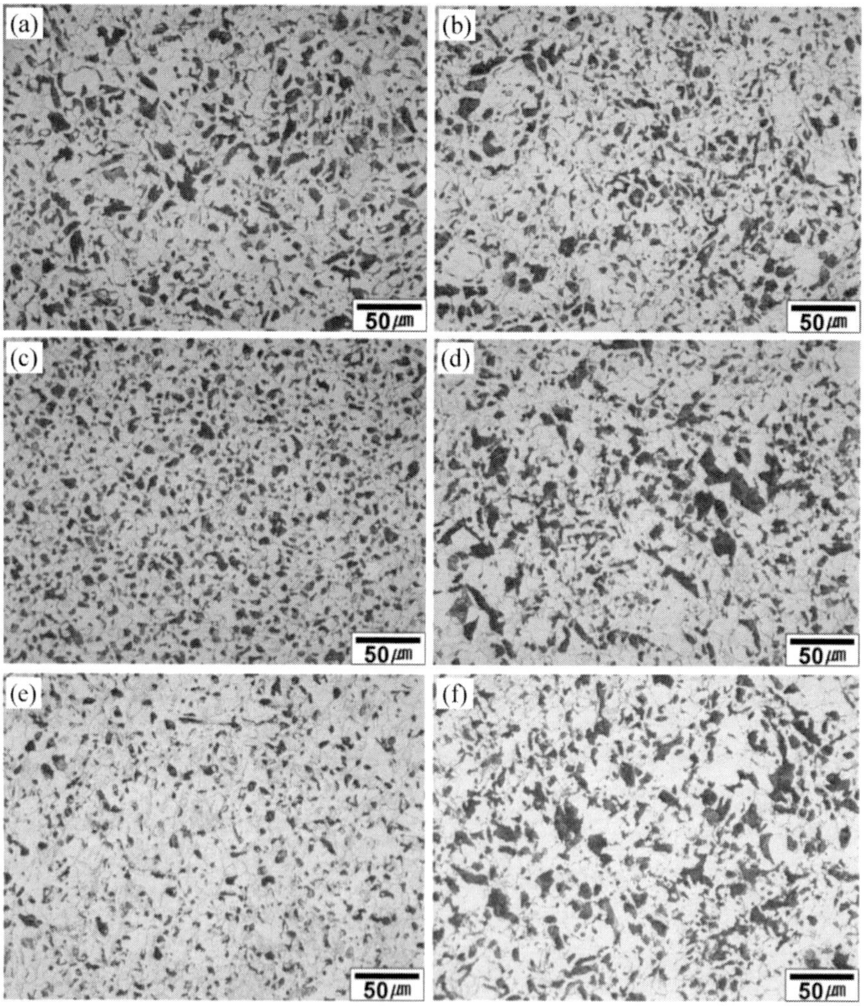

Figure 6.23. Microstructures of the steel at locations of (a) LS, (b) LC, (c) RS, (d) RC, (e) TS, and (f) TC.

Mechanical properties

Figure 6.24a shows the typical tensile curves of the specimens LC, RC and TC. The tensile properties determined from these curves are presented in Figure 6.24b.

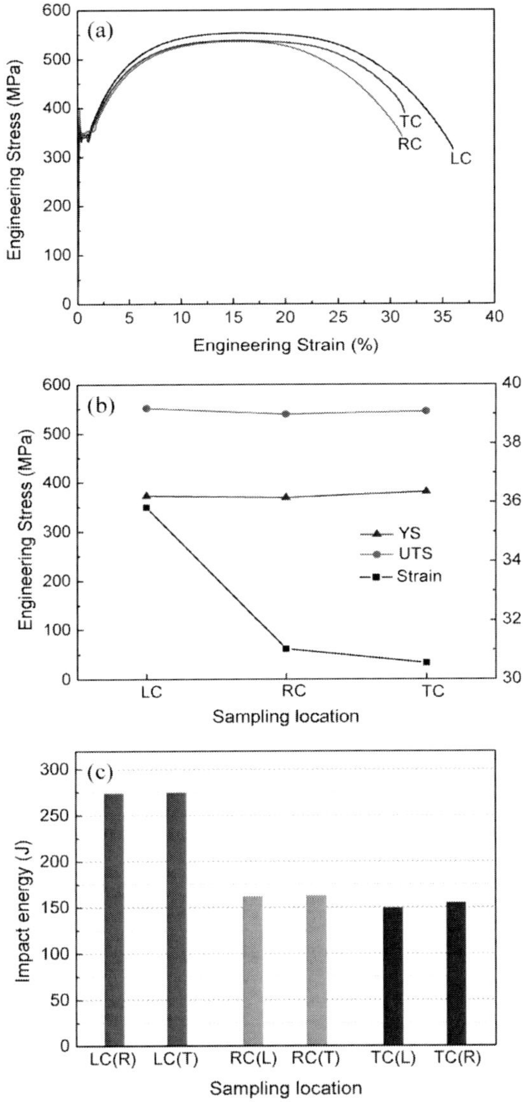

Figure 6.24. Mechanical properties of steel at different locations: (a) tensile curves, (b) tensile properties and (c) Charpy impact energies.

The room temperature Charpy impact energies are illustrated in Figure 6.24c. In Figure 4c, LC(R) means the notch direction in impact specimen LC is machined towards R. The results in Figure 6.24a indicate that the tensile behaviours of the tested specimens are similar. From Figure 6.24b, it can be found that sampling locations do not have noticeable effect on the tensile strength. Differently, the specimens RC and TC show lower strains in contrast to the specimen LC. The absorbed impact energies of the specimens prepared along the longitudinal direction are higher than that prepared along both the radius and thickness directions, as shown in Figure 6.24c. Moreover, notch direction slightly affects the impact energies.

In microalloyed steels, the tensile strength is closely related to the grain size. Fine-grained structure has a positive effect on the increment of tensile strength. According to Figure 6.23b, e and f, the mean average size in the specimen LC is similar to that in the specimens RC and TC. Therefore, no noticeable variation of tensile strength among specimens LC, RC and TC is observed. Directional properties in the ring-forged flange tend to vary between the longitudinal and transverse directions depending on the relative amounts of rolling work in each direction [15]. The variation in properties is due to directionality in the microstructure from forming operation. The specimens RC and TC are prepared perpendicular to the ring-forging direction, and lower values of strain and impact energy are expected relative to the specimen LC which is machined along the rolling direction. From the above results, it can be concluded that the designed heat treatment technology can be applied to the ring-forged heavy sectional steel flange to acquire uniformly distributed chemical compositions, microstructure and mechanical properties.

6.7. SUMMARY

This chapter has briefly introduced the purpose of heat treatment of steels and the parameters that are involved in heat treatment. The concepts, operations, advantages and applications of heat treatment processes including annealing, normalising, hardening and tempering that are commonly used in practice for steels especially for hypo-eutectoid steels have been described in order to provide a comparison of the various techniques.

The mechanical properties, especially impact toughness of microalloyed cast steel is very poor due to the coarse grains, inhomogeneous microstructure and serious carbon segregation. In order to modify the microstructure and achieve the desired mechanical properties of a microalloyed cast steel, several heat treatment processes have been designed with the purpose of finding the optimal heat treatment conditions correlated with the best combination of excellent toughness and tensile properties. Correct selection of the austenitising temperature, holding time and cooling method is very important to improve the mechanical properties. It has found that heat treatment at 1100°C for 2 h followed by furnace cooling results in the best combination of excellent impact toughness and tensile properties of the microalloyed cast steel.

Several microalloyed forging steels with different tungsten additions are heat treated in order to understand well the effects of tungsten content, heat treatment temperature and cooling method on the microstructure, impact toughness and tensile properties, and to obtain the optimal heat treatment conditions of these steels. The toughness of the heat-treated 0W steel suffered from air cooling is found to be better than that suffered from furnace cooling, while the heat-treated tungsten-containing steels subjected to furnace cooling show higher toughness levels than that subjected to air cooling. The optimal heat treatment temperatures of the 0W, 0.5W, 1W and 2W steels are 920, 900, 880 and 860°C respectively. Both the impact toughness and tensile properties of the microalloyed forging steels can be significantly enhanced after annealing heat treatment at the optimum temperatures followed by air cooling.

A real heavy sectional wind tower flange with dimensions of 3.7 m in inner diameter, 0.15 m in thickness and 0.19 m in height is produced by ring-forging. A heat treatment technology has been designed and successfully applied to this heavy sectional microalloyed steel to acquire uniformly distributed chemical compositions, microstructure and mechanical properties.

REFERENCES

[1] M.A. Grossmann, E.C. Bain, *Principles of Heat Treatment*, 5th ed., American Society for Metals, 1964.

[2] G. Krauss, *Steels: Processing, Structure, and Performance*, ASM International, 2005.

[3] J. Zhao, J.H. Lee, Y.W. Kim, Z. Jiang, C.S. Lee, Enhancing mechanical properties of a low-carbon microalloyed cast steel by controlled heat treatment, *Materials Science and Engineering A*, 559 (2013) 427-435.

[4] S.A. Hackney, G.J. Shiflet, Pearlite growth mechanism, *Acta Metallurgica*, 35 (1987) 1019-1028.

[5] EN 10025-3, *Hot Rolled Products of Structural Steels-Part 3: Technical Delivery Conditions for Normalized/Normalized Rolled Weldable Fine Grain Structure Steels*, British Standards Institution, London, 2004.
[6] J. Zhao, Z. Jiang, C.S. Lee, Effects of tungsten addition and heat treatment conditions on microstructure and mechanical properties of microalloyed forging steels, *Materials Science and Engineering A*, 562 (2013) 144-151.
[7] M. Srinivas, G. Malakondaiah, R.W. Armstrong, P. Rama Rao, Ductile fracture toughness of polycrystalline armco iron of varying grain size, *Acta Metallurgica et Materialia*, 39 (1991) 807-816.
[8] R.L. Klueh, D.J. Alexander, P.J. Maziasz, Bainitic chromium-tungsten steels with 3 Pct chromium, *Metallurgical and Materials Transactions A*, 28 (1997) 335-346.
[9] F.B. Pickering, *Transformation and Hardenability in Steels*, Climax Molybdenum Co., Ann Arbor, Michigan, 1967.
[10] J.P. Naylor, P.R. Krahe, The effect of the bainite packet size on toughness, *Metallurgical Transactions*, 5 (1974) 1699-1701.
[11] R.G. Davis, C.L. Magee, *Dual Phase and Cold Pressing Vanadium Steels in the Automobile Industry*, Vanitec, Berlin, 1978.
[12] S.R. Mediratta, V. Ramaswamy, V. Singh, P. Rama Rao, Microstructure-mechanical property correlations in dual phase steels, *Transactions of the Indian Institute of Metals*, 38 (1985) 350-372.
[13] A.H. Cottrell, *Dislocations and Plastic Flow in Crystals*, Oxford University Press, Oxford, England, 1953.
[14] A. Barbacki, The role of bainite in shaping mechanical properties of steels, *Journal of Materials Processing Technology*, 53 (1995) 57-63.
[15] E.G. Nisbett, Steel Forgings: Design, Production, Selection, Testing, and Application, *ASTM International*, 2005.

Chapter 7

EFFECT OF TUNGSTEN ON CORROSION RESISTANCE

ABSTRACT

Serious problems including steel part failures, plant shutdowns and pollution will be caused due to the occurrence of corrosion when microalloyed steels are used in a specific environment, e.g., sea water, leading to significant economic losses and/or expensive replacements even though the amount of steel destroyed is quite small. Therefore, the resistance to corrosion of microalloyed steels should be considered for assessing their performance before they are used in practice in combined with the environments getting involved. This chapter describes the effect of tungsten addition on the corrosion resistance of microalloyed steels. The categories of steel corrosion based on the mechanisms of damage caused are at first classified. Then the electrochemical corrosion reactions that may occur in microalloyed steels are briefly introduced. To help understand the corrosion process and control the occurrence of corrosion, the concepts of cathodic and anodic polarisations are also introduced. Next, the effects of tungsten on the anodic polarisation behaviour of microalloyed steels are investigated, and the evolution of oxide layer with tungsten is analysed. At last, the mechanism of improved corrosion resistance after tungsten addition is discussed. Tungsten is found to be beneficial for the increase of electrode potential and decease of current density. The oxide layer formed on the steel surface becomes more and more compact and uniform as the tungsten content is increased, and the number of microcracks in the oxide scale shows a decreasing trend with tungsten. These evidences indicate that the corrosion resistance of microalloyed steels has been improved after tungsten addition.

7.1. Categories of Steel Corrosion

Corrosion is typically defined as the deterioration of steel by electrochemical reaction with the environment, which leads to degradation of the properties and may cause functional damage of the steel parts or the whole system. There are many different specific environments depending upon how the steel is used. In the most general case, the environment getting involved of steel is a bulk aqueous solution. For atmospheric corrosion, the aqueous solution is a condensed thin-layer rather than a bulk solution, but the overall principles are the same. Corrosion of steels can be classified into different categories based on the mechanisms of the damage caused.

- **General corrosion.** General corrosion, which is also called uniform corrosion, is the most common form of corrosion and it represents the greatest steel loss on a tonnage basis. It occurs over the majority of the surface of steel at a steady and often predictable rate. General corrosion is a uniform thinning of steel without localised attack, and corrosion does not penetrate very deep inside. It could lead to failure through lowering the mechanical strength of steel parts or by reducing wall thickness until leaking results. Rust formed on low alloyed steels is a very familiar example of general corrosion. General corrosion, however, is a less dangerous form of corrosion because it is predictable and efficient protection methods are available to reduce the corrosion rate to an acceptable level.
- **Galvanic corrosion.** Galvanic corrosion is an electrochemical process in which steel contacts with another conducting material in an electrolyte. Dissimilar metals and alloys may have different electrode potentials, and when two or more are coupled in an electrolyte, one metal will act as anode and corrodes faster than it would all by itself, while the other becomes cathode and corrodes slower than it would alone. Galvanic corrosion is a common type of corrosion found in marine environment due to the salt water's effectiveness as an electrolyte. For example, two dissimilar metals such as steel and aluminium used in marine structures touching and an electrolyte such as sea water or even rain water may form a complete electrical circuit for the occurrence of galvanic corrosion. The extent of galvanic corrosion depends on a variety of factors including the ratio of the areas of each metal exposed to the environment, the potential difference between the metals, the electrical resistivity of the joint,

and the conductivity of the environment. Prevention of galvanic corrosion should be based on the guarantee that one or more of these factors do not exist.

- **Pitting corrosion.** It is a localised form of corrosion by which small "pits" or "holes" are produced in the steel. Pitting corrosion is one of the most damaging forms of corrosion because it is difficult to detect, predict and design against. It commonly occurs on passive metals such as stainless steels when the ultra-thin passive film, such as oxide film, is chemically or mechanically damaged and does not immediately re-passivate. The resulting pits can become wide and shallow or narrow and deep which can rapidly perforate the wall thickness of a metal. Therefore, a small, narrow pit with minimal overall steel loss can lead to the failure of an entire engineering system. To prevent the occurrence of pitting corrosion, the passive film formed on the steel must be continuous and it must be able to reform rapidly by reaction between the steel and the environment when the film cracks, ruptures or become damaged through any mechanical or chemical process.
- **Crevice corrosion.** It is a form of localised corrosion usually associated with the presence of stagnant solution in occluded interstices, beneath deposits and seals, or in crevices, where the environment in these regions do not freely mix with the bulk environment. Examples of crevices include gaps and contact areas between steel parts, under gaskets or seals, inside cracks and seams, spaces filled with deposits and under sludge piles, and under bolds and rivet heads, etc. Crevice corrosion is frequently observed in stainless steels when they are exposed to an environment that contains halide irons especially chloride. The corrosion resistance of a stainless steel depends upon the presence of an ultra-thin protective passive film on its surface, but such film may become unstable and be broken down under certain conditions, for example in the presence of high concentrations of Cl^- and H^+ ions generated from halide solutions. Such corrosion attack can occur at rates that are orders of magnitude greater than rates occurred outside the crevice once this attack is initiated. Failure through leaking, mechanical failure, freezing and initiation of cracks that propagate failure of steels joints could be caused when crevice corrosion occurs.
- **Stress corrosion.** As the name implies, stress corrosion involves the interaction of mechanical stress and corrosion. It occurs when steel

exists in a relatively inert environment but corrodes due to an applied stress. Stress corrosion is evidenced when the steel strength loss resulting from the combined stress and corrosion is greater than the effects of stress and corrosion acting separately. The magnitude of the combined effect is a measure of the susceptibility of the steel to stress corrosion. The result of stress corrosion is usually accompanied by an appearance of cracks in the steel propagating either transgranularly or intergranularly. If the stress applied is constant during the period of cracks growth, steel failure will be caused as the size of cracks reaches a critical value. Not all types of stresses facilitate stress corrosion. Resultant cracking due to stress corrosion is often caused by residual stresses generated during forging, welding or assembly, etc. For reducing the risk of stress corrosion cracking, it is especially important to minimise the generation of residual stresses and avoid any mechanical tensile stress concentration which generally occurs at sharp edges and notches.

- **Intergranular corrosion.** Intergranular corrosion, also known as intergranular attack, is a form of localised attack in the vicinity of the grain boundaries, or immediately adjacent to grain boundaries of steel, while the interiors of the grains remain un-corroded or rarely corroded. It occurs when the grain boundaries in a steel form an anode and the interior of the grain acts as a cathode. This form of corrosion is usually associated with the presence of impurities at grain boundaries, local enrichment or depletion of one or more alloying elements, or specific phases precipitated at the grain boundaries. For example, grain boundaries precipitation, notably chromium carbides in stainless steels, is a well-recognised mechanism of intergranular corrosion. The attack usually progresses along a narrow path of grain boundary and, entire grains may be dislodged due to complete deterioration of their boundaries in a severe case of intergranular corrosion. The surface of the steel can appear un-attacked when intergranular corrosion takes place, the mechanical properties of the steel, however, will deteriorate slowly or rapidly until the steel failure under applied stress or until individual grains starts to fall out and the steel falls apart.
- **Dealloying corrosion.** Dealloying corrosion is also called selective corrosion or selective leaching. It is the removal of an element from a steel by corrosion. Dealloying corrosion is observed in a steel in which the less noble element is removed from the surface by a

microscopic-scale galvanic corrosion mechanism due to the differences in chemical activities among different elements. This selective leaching process of an alloying element may cause the surface being dealloyed and in some cases the dealloyed layer may propagate into the inner of the steel part. The steel in the affected area of dealloying corrosion becomes porous and loses much of its strength, hardness and ductility, and failure may occur suddenly and unexpectedly.

- **Concentration cell corrosion.** It occurs when two or more areas of a steel surface are in contact with different concentrations of the same electrolyte. The extent of this corrosion is proportionate to the difference in concentrations at contact points. This difference in concentrations, which can be due to non-uniform deposits or fouling on the surfaces, or, more commonly, built in features which create significant difference in environmental concentrations, can lead to some areas of the steel acquiring different electric potentials. Concentration cell corrosion is most prevalent in the presence of oxygen. When pure oxygen comes into contact with a wet steel surface, corrosion action is enabled. Steel parts that are covered by scale will corrode fast because the contact with oxygen for these parts is restricted. Concentrated pitting can be caused due to this cumulative reaction. Concentration cell corrosion is important because such corrosive damage is commonly observed in critical applications, such as pipe lines buried in soil, due to the variable characteristics of the soil.
- **Erosion corrosion.** Erosion corrosion is the acceleration in the rate of corrosion attack due to the relative movement between a corrosive fluid and a steel surface. It typically occurs in pipe bends (elbows), tube constrictions and other structures that alter flow direction or velocity. Erosion corrosion can also result from poor workmanship. For example, when burrs in the tubes are not removed during installation, these inner burrs may cause localised turbulence and hinder the smooth flow of the fluid. This leads to high rates of corrosion in the tubes. This form of corrosion involves removing any protective film or metal oxide from the steel surface due to the continuous flow of fluid. Once the steel surface is exposed, it is attacked by the corrosive media and eroded away by the fluid friction. If the passive layer of metal oxide cannot be regenerated quickly enough, significant damage may occur.

In practice, a steel is usually subjected to more than one of the aforementioned corrosion failure modes. The consequences of corrosion are many and varied and the effects of these on the safe, reliable and efficient operation of equipment and structure are often more serious than the simple loss of a mass of steel. Steel part failures, plant shutdowns and pollutional problems will be caused due to the occurrence of corrosion, leading to significant economic losses and/or expensive replacements even though the amount of steel destroyed is quite small. For preventing the occurrence or reducing the magnitude of corrosion of steels, the factors that can tend to accelerate the actions of steel corrosion should be avoided. Details about the control of various types of steel corrosion can be found in Refs. [1-5].

7.2. Electrochemical Corrosion Reactions

Steels may corrode in a variety of gaseous and aqueous environments. In this chapter, focus is put on the most common electrochemical corrosion of steels in aqueous solution. If a steel is immersed into an aqueous solution, the oxide film formed on the steel surface tends to dissolve. When the oxide-free surface of a steel become exposed to the solution, positively charged iron ions may pass from the steel into the solution, and at the same time, leave electrons behind on the steel:

$$\mathrm{Fe} \rightarrow \mathrm{Fe}^{2+} + 2e^{-} \tag{7.1}$$

In acid solutions, electrons released will react with hydrogen ions that adsorbed on the steel surface from the solution to produce hydrogen gas:

$$2\mathrm{H}^{+} + 2e^{-} \rightarrow \mathrm{H}_2 \tag{7.2}$$

In neural solutions, the concentration of hydrogen ions is too low to allow reaction (7.2) to proceed at a significant rate. Alternatively, electrons will preferentially react with oxygen molecules that adsorbed on the steel surface from air dissolved in the solution to produce hydroxyl ions:

$$\mathrm{O}_2 + 2\mathrm{H}_2\mathrm{O} + 4e^{-} \rightarrow 4\mathrm{OH}^{-} \tag{7.3}$$

Reaction (7.1) is called oxidation reaction of corrosion by which steel dissolution occurs through the loss of electrons. Correspondingly, reactions (7.2) and (7.3) are called reduction reactions since they both involve gain in electrons. An electrode at which an oxidation reaction occurs is termed as anode, whereas the area of a corroding steel where the reduction reaction of hydrogen ions or oxygen occurs is called cathode. The cathode acts as a receptor for the electrons during electrochemical corrosion reactions. A larger cathodic area may draw more electrons from the anode, leading to more steel dissolution at the anode. In the case of a very small anodic area and a very large cathodic area, almost all the steel in the anodic area will dissolve in order to satisfy the electron demand from the chathode, which may result in the occurrence of localised and/or pitting corrosion. Conversely, a general corrosion will be caused if the anodic area is large enough relative to the cathodic area due to the decrease in corrosion rate under the same environment and condition.

7.3. Electrochemical Polarisation

The corrosion reaction involves the transfer of electrons and ions between the steel and the solution, which causes a flow of current. The change in electrode potential due to the flow of current is referred to as polarisation, which may be preferential to either the cathodic or anodic reactions. Cathodic polarisation is the displacement of the electrode potential in the negative direction so that the electrode acts more cathodic. Therefore, cathodic polarisation is negative by definition. Anodic polarisation is the opposite process. Figure 7.1 plots schematically the juxtaposition of both cathodic and anodic polarisation curves of a steel corroding in an acid. In the figure, β_a and β_c are the Tafel slopes of the anodic and cathodic polarisation curves, respectively, which are determined by $\beta = dE/di$. The Tafel slope β is related to the kinetics of the corrosion reaction. E_{corr} is the corrosion potential at which the rate of anodic reaction is equal to the rate of cathodic reduction reaction of hydrogen ions or oxygen. i_{corr}, the corrosion current, is determined by extrapolating the straight line portion, or Tafel regions, of either the anodic or cathodic polarisation curve to the corrosion potential E_{corr}, and the intersection there gives the value of i_{corr}. The extent of polarisation can help predict the type and severity of corrosion. Corrosion decreases with the increase of polarisation. An understanding of the effect of environmental changes on polarisation can aid in controlling the occurrence of corrosion. A wealth of relevant information addressing the electrochemical polarisation of steels is available in Refs [3,6,7].

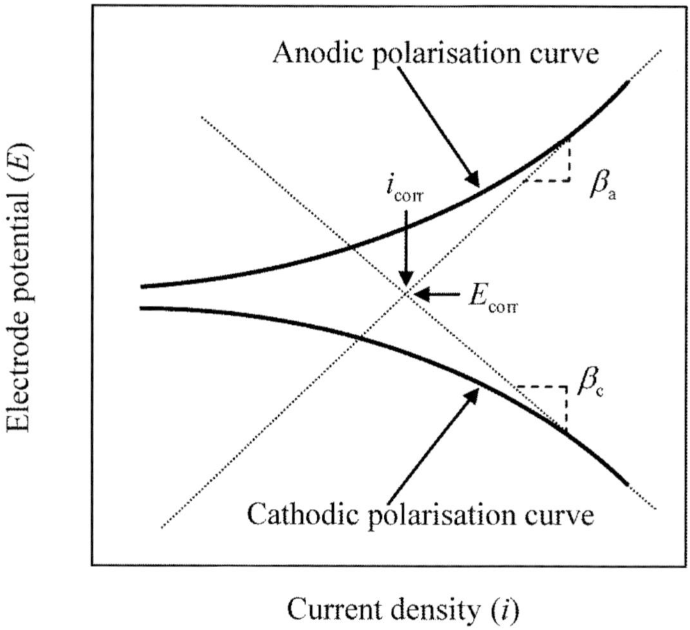

Figure 7.1. Schematic anodic and cathodic polarisation curves of a steel corroding in an acid.

7.3.1. Cathodic Polarisation

In most corrosion reactions, cathodic polarisation is more predominant because the degree of cathodic polarisation and its effectiveness varies with steels [7]. For example, some stainless steel grades are very noble and may demonstrate excellent corrosion resistance to sea water. When a less noble steel, such as microalloyed steel, is coupled to such a stainless steel, galvanic corrosion is usually accelerated much less than that would be anticipated because stainless steel polarises readily and quickly in sea water, thereby significantly reducing corrosion rate.

Figure 7.2 shows a schematic cathodic polarisation curve (ABCD) that a steel may experience in an electrochemical corrosion process. In the cathodic polarisation curve, the potential is varied from point A in a negative direction to point D. The open circuit potential is located at point A. At this potential, the sum of the anodic and cathodic reaction rates on the electrode surface is zero.

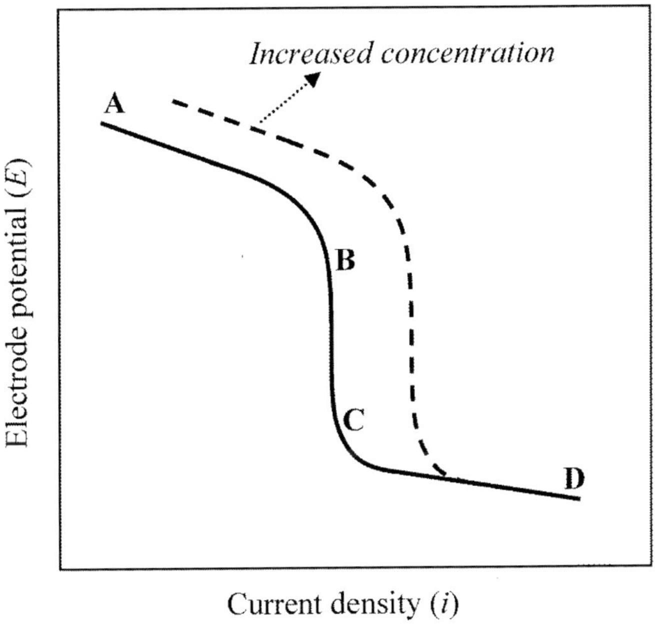

Figure 7.2. Schematic illustration of a cathodic polarisation curve that a steel may experience as well as the effect of increasing concentration on the curve.

During the cathodic reduction reactions, either hydrogen ions or oxygen molecules must be transported from the bulk solution to the steel/solution interface. As the potential of the steel is made more negative in the Tafel region (Figure 7.1), the rate of reduction of hydrogen irons or oxygen molecules may increase to such an extent that it exceeds the rate at which these species can be transported to the steel surface. Then the rate of the reducible species will dominate the rate of reduction. The cathodic current density reaches a limiting value equivalent to the rate of transport, and is independent of potential, as shown at the region of BC in Figure 7.2. In this region, further decreases in the applied potential result in no change in the reaction rate, and hence the current density. The value of the limiting current density increases with the increase of the concentration of the reducible species, as indicated by the dash line in Figure 7.2. When the potential becomes sufficiently negative, the rate of reduction of water to produce hydrogen may exceed the diffusion limited rates of reduction of hydrogen ions or oxygen molecules, then the current density will increases due to the reduction of water, as along CD shown in Figure 7.2.

7.3.2. Anodic Polarisation

For practical purposes, anodic polarisation is often used to understand the steel systems in various application environments. Anodic polarisation curves are extremely useful in the investigation of the active-passive behaviour that some steel grads may exhibit. Stainless steels are commonly subjected to this type of behaviour. Active-passive behaviour can be well observed by potentiodynamic polarisation, which is due to the reaction of the steel surface with solution to form a passive oxide film. In the potentiodynamic polarisation method, the potential of the working electrode continuously increases at a very slow constant scanning rate with a potentiostat. The potentiodynamic method of polarisation can follow more closely the behaviour of steels during the formation and breakdown of passive films, where the behaviour depends primarily on the potential of the steel, and significant changes in currents can be observed with the continuous increase in potential. An anodic polarisation curve exhibiting distinguished active-passive feature is schematically shown in Figure 7.3.

As shown in Figure 7.3, the curve starts from point A and progresses in the positive potential direction until termination at point E. The open circuit potential is located at point A. As with cathodic polarisation, the open circuit potential represents the potential at which the sum of the anodic and cathodic reactions occurring on the electrode surface is zero. With an increase of the potential, the anodic polarisation steel shows a distinct behaviour, and a number of notable features can be observed on the polarisation curve. At low potentials, the current density increases gradually from point A as expected with potential in the *active* region. In this region, steel oxidation is the dominant reaction occurring. With the progress of polarisation, oxide gradually spread over the steel surface to form a thin passive oxide film. In stainless steels this film is generally composed of various chromium and/or nickel oxides which exhibit substantially different electrochemical characteristics as compared to the underlying steel substrate. The passive oxide film begins to form near point B of the polarisation curve AB, and the corresponding potential and current density at this peak point are known as critical passivation potential, E_{cp}, and critical passivation current density, i_{cc}. The value of the critical passivation current density provides a measure of the ease of passivation, i.e., the smaller the current, the easier is passivation. When a passive film is formed on the steel surface, the steel then moves to the *passive* region.

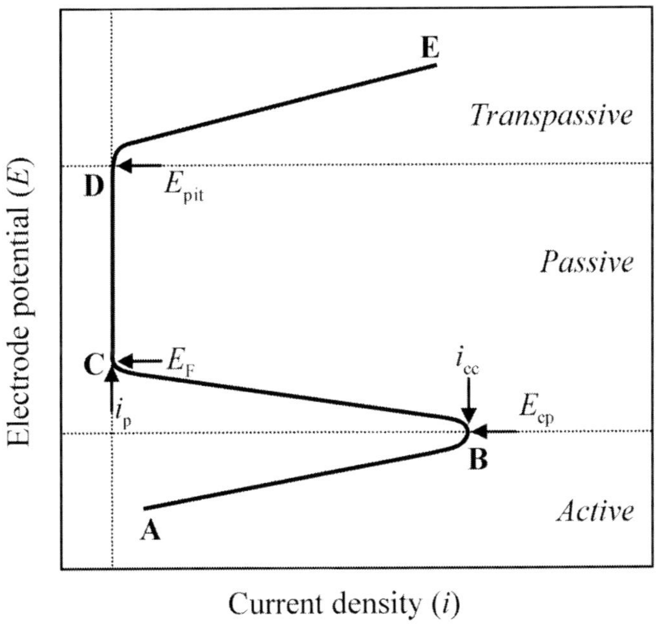

Figure 7.3. Schematic illustration of an anodic polarisation curve of a steel exhibiting active-passive behaviour.

Above the potential of E_{cp}, the passive film becomes stable and the corrosion rate decreases significantly. The passage of iron ions through the oxide film takes place very slowly and so the current induced by the transfer of iron ions becomes very small when the surface is completely covered with a passive oxide film. As the thickness of oxide film increases to an equilibrium value, the current of iron ions passing through the film become independent of potential. The potential where the anodic current first drops to its minimum is named Flade potential, E_{F}, in honour of the scientist Flade [8] who was the first to observe passivity of iron in sulphuric acid in the year 1911, as shown at point C in Figure 7.3. The value of E_{F} indicates the most negative limit of stability of the passive state. The current density required to maintain the passive state, i_{p}, is a measure of the protectiveness of the film. Usually, the value of i_{p} is several order of magnitudes lower than that of i_{cc}. As the potential is increased further, i_{p} remains reasonably unchanged until eventually it starts to increase with potential beyond point D, and then the steel enters the *transpassive* region, as shown in Figure 7.3. The corrosion rate of the steel is not zero in the passive state region CD, but is very small and maintains a constant value.

The increase in current density beyond point D with potential is due to the localised breakdown of the protective oxide film. The breakdown of the passive oxide film can be caused either mechanically or electrochemically. While the underlying steel substrate is freshly exposed, an increase in the anodic current density will then be induced. Continuous breakdown of the passive oxide film will lead to the formation of pits. The potential at which the initiation of pitting occurs is called the pitting potential, E_{pit}, beyond which the steel enters the *transpassive* region, and the current density increases gradually with potential, as along DE shown in Figure 7.3.

7.4. Effect of Tungsten on Polarisation Behaviour

7.4.1. Polarisation Test

To study the effect of tungsten addition on the corrosion resistance of microalloyed steels, the anodic polarisation behaviour of the steels was analysed. A computer-controlled Potentionstat/Galvanostat instrument, as shown in Figure 7.4, was employed to obtain the data used for the experimental polarisation curves. To measure the electrode potential of the steels, potentiodynamic anodic polarisation tests were conducted in a 3.5% NaCl (in mass percent) solution at room temperature.

Figure 7.4. Potentionstat/Galvanostat instrument used for obtaining the anodic polarisation data.

This solution was determined based on the sea water environment in which microalloyed steels might be applied. To study the effect of the concentration of NaCl solution on the polarisation behaviour, a milder solution with 1% NaCl (in mass percent) was also employed in the anodic polarisation tests. The surfaces of all the specimens (diameter of 15 mm and height of 5 mm) were mechanically polished with #2400 SiC sand paper, degreased with acetone, rinsed with bidistilled water and dried between two filter papers so that the conditions prior to exposure were the same for each steel specimen. The corrosion area of the specimen was specified as 1 cm^2. The tests were conducted at a continuously varying potential range of 0 to +300 mV with a scanning rate of 0.5 $mV{\cdot}s^{-1}$.

7.4.2. Polarisation Behaviour

Effect of NaCl concentration

Figure 7.5 compares the potentiodynamic anodic polarisation curves of the 0W steel corroded in both 1% and 3.5% NaCl solutions. In the figure, V_{SCE} is the potential measured versus the saturated calomel electrode (SCE). SCE is composed of mercury in contact with a saturated solution of calomel, and is a commonly used reference electrode based on the reaction between elemental mercury and mercury chloride. As indicated in Figure 7.5 [9], no passive region appears in the polarisation curves. The increase of concentration of NaCl in the solution leads to an increase of current density at certain definite potential denoting the destruction of the oxide film. The lower the concentration of NaCl, the greater extent of the electrode potential shifts to the positive direction, which indicates that severity of corrosion of the 0W steel is enhanced as the concentration of NaCl in the corroding solution is increased from 1% to 3.5%.

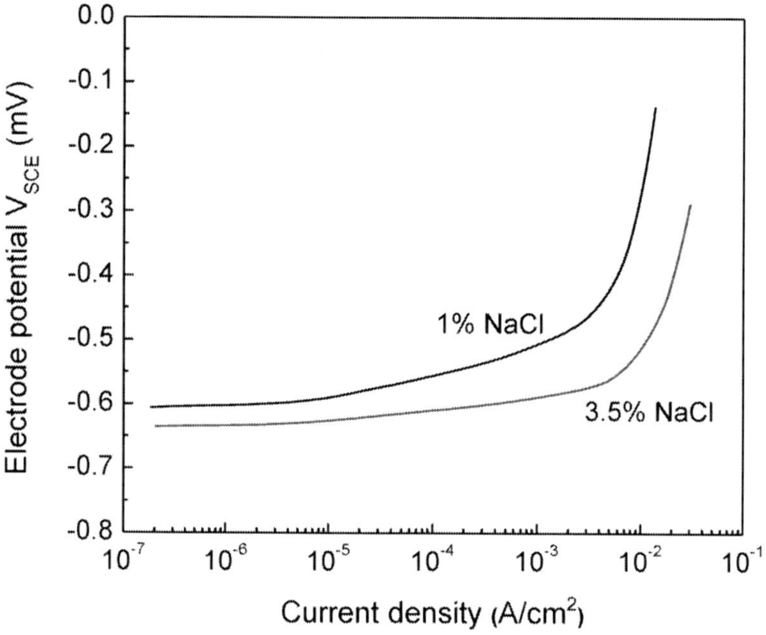

Figure 7.5. Anodic polarisation curves of the 0W steel corroded in both 1% and 3.5% NaCl solutions [9].

When the NaCl concentration is increased, the number of sodium and chloride ions dissociated will increase. These ions act as charge carriers and increase the conductivity of the solution. Because corrosion is an electrochemical process, the increased conductivity may lead to an increase in the rate at which the steel is oxidised, and consequently cause acceleration in the corrosion. Therefore, the corrosion of the 0W steel can be accelerated by the increase of NaCl concentration.

Effect of tungsten content

Figure 7.6a and b plot the electrode potential against current density for the potentiodynamic anodic polarisation of the 0W, 0.1W and 0.5W steels in 1% and 3.5% NaCl solutions, respectively [9].

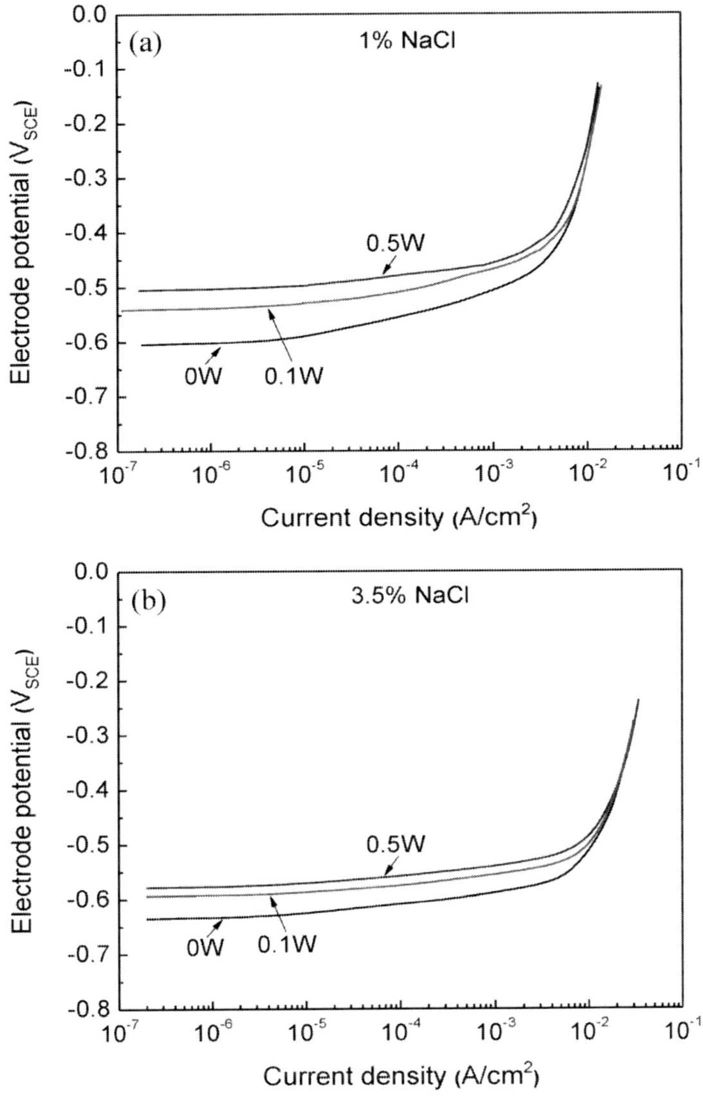

Figure 7.6. Anodic polarisation curves of the 0W, 0.1W and 0.5W steels in (a) 1% and (b) 3.5% NaCl solutions [9].

It can be seen that the shape of the polarisation curve of the 0W steel is similar to that of both the 0.1W and 0.5W steels. The addition of tungsten acts to cause the electrode potential more positive and the current density more negative with respect to the 0W steel. It indicates that the addition of tungsten can enhance the corrosion resistance, and the severity of corrosion shows a decreasing trend with the increase of tungsten content.

7.5. Effect of Tungsten on Oxide Layer

Figure 7.7 shows the SEM micrographs of the corroded surfaces of the 0W and 0.1W steels in 3.5% NaCl solution at a low magnification of ×50. It can be seen that the area of oxide layer formed on the steel surface increases as tungsten is added, indicating the positive role of tungsten addition in the facilitation of oxide layer formation on microalloyed steels.

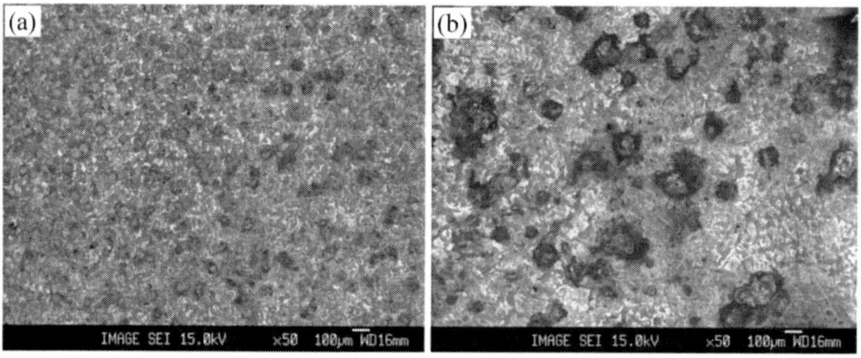

Figure 7.7. Lower magnification SEM micrographs of the corroded surfaces of the (a) 0W and (b) 0.1W steels in 3.5% NaCl solution.

Figure 7.8 shows the SEM micrographs of the corroded surfaces of the 0W and 0.1W steels in 3.5% NaCl solution at magnifications of ×1,000 and ×10,000 [9]. As indicated in Figure 7.8a and b, small-sized oxide scale distributes separately on the 0W steel, while massive oxide scale appears on the 0.1W steel. Observations on the oxide scales at a higher magnification of ×10,000 show that a large number of microcracks exist in the oxide scale formed on the 0W steel, as shown in Figure 7.8c. Differently, a remarkable decrease in the quantity of microcracks in the oxide scale is observed after 0.1% tungsten addition, as shown in Figure 7.8d.

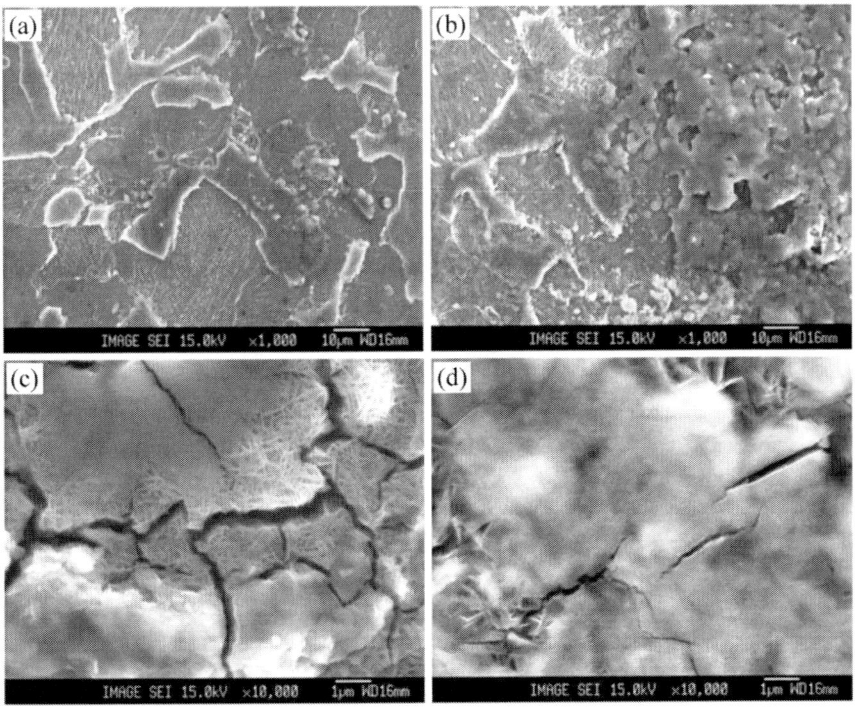

Figure 7.8. Higher magnification SEM micrographs of the corroded surfaces of the (a,c) 0W and (b,d) 0.1W steels in 3.5% NaCl solution [9].

With further increasing the tungsten content from 0.1% to 0.5% and much higher value, the morphologies of the corroded surfaces have significantly changed. As shown in Figure 7.9a, the oxide layer formed on the 0.5W steel consists of short-bar shaped oxide scales which cover almost the whole corroded surface. As the tungsten content is increased to 2%, a compact and uniform oxide layer forms, and no exposed steel surface is observed, as shown in Figure 7.9b.

Higher magnification observations on the oxide scales show that only a very small quantity of microcracks is found in the oxide scale formed on the 0.5W steel, while no any microcrack is observed in the oxide scale formed on the 2W steel, as indicated in Figure 7.9c and d. The morphology of the oxide layer formed on the steel surface reflects the corrosion resistance. Generally, a compact and uniformly distributed corrosion oxide layer which contains few or no microcracks corresponds to a relatively high corrosion resistance of the steel.

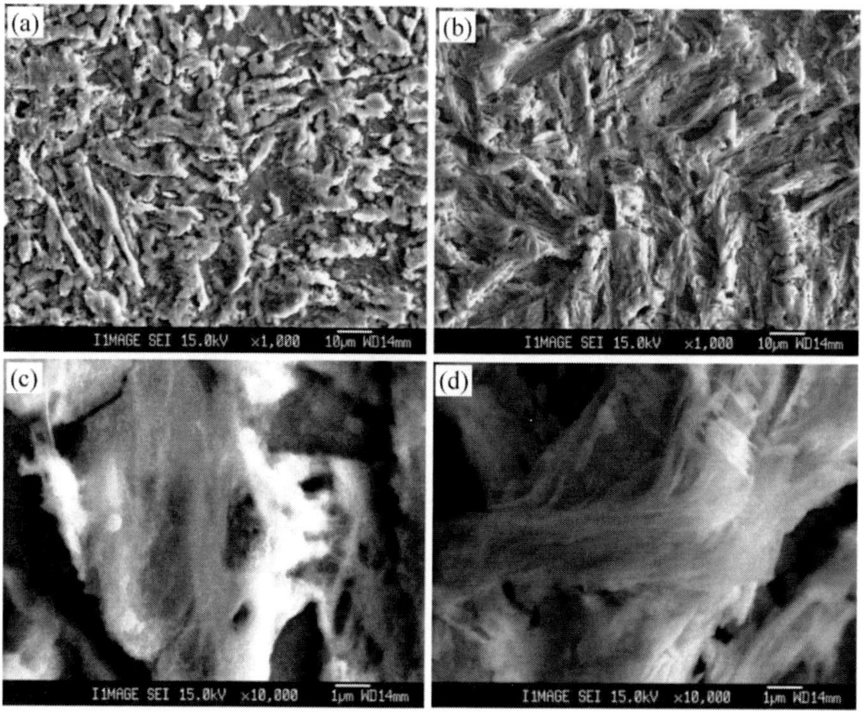

Figure 7.9. SEM micrographs of the corroded surfaces of the (a,c) 0.5W and (b,d) 2W steels in 3.5% NaCl solution.

7.6. Mechanism of Improved Corrosion Resistance after Tungsten Addition

Tungsten was initially added in 20Cr-25Ni-MoCu steels to enhance their abrasion resistance [10], progressively it has approved to be beneficial to corrosion resistance of several grades of steels/irons including stainless steels [11,12], microalloyed steels [9], cast irons [13,14], and low alloy steels [15], etc.

Anodic polarisation behaviour of the tungsten-containing steels is affected by the tungsten content. For example, for tungsten-containing Fe-29Cr ferritic stainless steels, no passive region was observed even for the steel containing up to 2 wt.% tungsten [16]. Therefore, there should be a critical tungsten content to cause the occurrence of passivation in steels. It explains why the anodic polarisation curves of the 0W, 0.1W and 0.5W steels do not show passivation characteristics, as shown in Figure 7.6.

It has been generally accepted that the addition of tungsten in steels can improve the electrode potential and decrease the corrosion rate. The improved corrosion resistance of steels after tungsten addition is realised by its role in the help of the formation of a passive oxide layer on the steel surface. Chen and Wu [14] found that the formation of insoluble oxide WO_3 could enhance the stability of the oxide layer formed on the iron surface, and thus improved the corrosion resistance after alloying of tungsten in iron or addition of WO_4^{2-} ions in solution. Gribaudo and Rameau [17] claimed that tungsten added in the as-cast 25Cr-20Ni austenitic stainless steel promoted $M_{23}C_6$ carbides precipitation and consequently lowered the occurrence of local corrosion. The work of Bui et al. [18] on 16Cr-14Ni-*x*W (*x*=0-12) austenitic stainless steels showed that the electrode potential of the steels increased with the content of tungsten in NaCl solution, and claimed this was associated with the direct interaction of tungsten with water to form insoluble WO_3 in the passive film, followed by the interaction of WO_3 with other oxides to form a complex oxide film. The oxide film can be divided into two layers, namely inner layer and outer layer. Irhzo et al. [9] found that the formation of WO_3 occurred in the inner layer of the passive film due to the fact that this layer is related to the existence of a highly resistive ionic barrier. Concerning the outer layer, the WO_3 species is thought to trap the negative charge carriers in the film to become WO_4^{2-}, which may constitute the charges having a large time constant in transient currents due to the less mobility.

In addition, the formation of WO_4^{2-} ions may cause a decrease in the positive charge density in the passive film, leading to a reduction in the migration rate of ions and decrease in the attractive force between the positive ions and the anions from the electrolytic solution, such as chloride ions. Chloride ions can form soluble complex metallic chlorides upon adsorption onto the passive film and to initiate the pitting corrosion. In duplex stainless steels, tungsten is usually added in combined with molybdenum at a specific ration to improve the corrosion resistance. Even though similar chemical behaviours of tungsten and molybdenum are expected due to the fact that they belong to the same family of VIB in a periodic table, the effects of them on the corrosion resistance however are different. Kim and Kwon [19] pointed out that resistance to localised corrosion of duplex stainless steels was improved with a partial substitution of molybdenum by tungsten in alloying composition. Park and Kwon [11] confirmed that the corrosion resistance of

25%Cr duplex stainless steels could be most improved by alloying molybdenum and tungsten at a specific ratio such as 1 to 2.

Noda et al. [20] reported that the addition of tungsten was effective in the decrease of corrosion rate of iron, despite their anion selective permeability of rust film. Itagaki et al. [7.15] studied the effects of tungsten addition on the electrochemical behaviour of low alloy steels in NaCl aqueous solution. They found that tungsten existed as a WO_4 compound in the oxide film on the Fe-W low alloy steel. The existence of such compound complicated the structure of the oxide film and thus lowered the ion permeability rate inside the oxide film, and consequently the corrosion rate of the steel was decreased.

The beneficial effect of tungsten alloying on the corrosion resistance can also be interpreted by other possible mechanisms [9,21-23]: (i) stabilisation of the oxide film; (ii) formation of defect free oxide film in which ionic conductivity if very low; and (iii) decrease of the activity of micropores or weak points.

7.7. SUMMARY

In this chapter, the categories of steel corrosion, the electrochemical corrosion reactions and the electrochemical polarisation are introduced with the purpose of understanding well the corrosion process of steels. The effects of tungsten addition on the anodic polarisation behaviour and the features of the oxide layer are analysed. The mechanism of tungsten-induced improvement in the corrosion resistance is discussed.

No passive region is found in all the anodic polarisation curves due to the low content of tungsten in the microalloyed steels. The increase of concentration of NaCl in the aqueous solution leads to an increase in current density and a decrease in electrode potential. The shape of the anodic polarisation curve of the 0W steel is similar to that of both the 0.1W and 0.5W steels. The addition of tungsten acts to cause the electrode potential more positive and the current density more negative with respect to the 0W steel.

The area of oxide layer formed on the 0W steel surface increases as tungsten is added. Small-sized oxide scale distributes separately on the 0W steel, and the oxide scale contains a large number of microcracks. Massive oxide scale appears on the 0.1W steel and the quantity of microcracks in the oxide scale is significantly reduced as compared to the 0W steel. When the tungsten content is increased to 0.5%, the oxide layer consists of short-bar shaped oxide scales which cover almost the whole corroded surface. Only a very small quantity of microcracks is found in the oxide scale formed on the 0.5W steel. As the tungsten content is increased to 2%, a compact and uniform oxide layer forms, and no exposed steel surface is observed. Higher magnification observations on the oxide scales show that no any microcrack is observed in the oxide scale formed on the 2W steel. The improved corrosion resistance of microalloyed steels after tungsten addition is associated with the formation of a compact and uniformly distributed corrosion oxide layer which contains few or no microcracks.

REFERENCES

[1] Rudy J.M. Konings (Ed.), *Comprehensive Nuclear Materials*, Elsevier, 2012.

[2] Tony J.A. Richardson (Ed.), *Shreir's Corrosion*, Elsevier, 2010.

[3] R. Winston Revie, H.H. Uhlig, *Corrosion and Corrosion Control*, 4th ed., John Wiley & Sons, Inc., 2008.

[4] C. Dinsdale, *Prevention of Iron and Steel Corrosion - Processes and Published Specifications*, Louis Cassier Co., London, 1948.

[5] A. Sang, *The Corrosion of Iron and Steel*, McGraw-Hill Book Company, New York, 1910.

[6] P.A. Schweitzer, *Encyclopedia of Corrosion Technology*, 2nd ed., Marcel Dekker, Inc. 2004.

[7] J.R. Davis, *Corrosion: Understanding the Basics*, ASM International, 2000.

[8] F. Flade, Contributions to the knowledge of passivity, *Zeitsch Physikal Chemistry*, 76 (1911) 513-546.

[9] J. Zhao, Y.W. Kim, J.H. Lee, J.M. Lee, H.S. Chang, C.S. Lee, Effect of tungsten addition on the mechanical properties and corrosion resistance of S355NL forging steel, *Metals and Materials International*, 18 (2012) 217-223.

[10] A. Irhzo, Y. Segui, N. Bui, F. Dabosi, The role of alloyed tungsten on the conductivity of stainless steel passive layers, *Corrosion Science*, 26 (1986) 769-780.

[11] C.J. Park, M.K. Ahn, H.S. Kwon, Influences of Mo substitution by W on the precipitation kinetics of secondary phases and the associated localized corrosion and embrittlement in 29% Cr ferritic stainless steels, *Materials Science and Engineering A*, 418 (2006) 211-217.

[12] C.J. Park, H.S. Kwon, Effects of aging at 475℃ on corrosion properties of tungsten-containing duplex stainless steels, *Corrosion Science*, 44 (2002) 2817-2830.

[13] S.H. Mousavi Anijdan, A. Bahrami, N. Varahram, P. Davami, Effects of tungsten on erosion–corrosion behavior of high chromium white cast iron, *Materials Science and Engineering A*, 454-455 (2007) 623-628.

[14] J. Chen, J.K. Wu, The improved passivation of iron induced by additions of tungsten. *Corrosion Science*, 30 (1990) 53-58.

[15] M. Itagaki, R. Nozue, K. Watanabe, H. Katayama, K. Noda, Electrochemical impedance of thin rust film of low-alloy steels, *Corrosion Science*, 46 (2004) 1301-1310.

[16] M.K. Ahn, H.S. Kwon, H.M. Lee, Quantitative comparison of the influences of tungsten and molybdenum on the passivity of Fe-29Cr ferritic stainless steels, *Corrosion Science*, 40 (1998) 307-322.

[17] L.M. Gribaudo, J.J. Rameau, General and local corrosion of as-cast 25Cr-20Ni steel in molten sulphates: the influence of Si and W additions, *Corrosion Science*, 24 (1984) 291-303.

[18] N. Bui, A. Irhzo, F. Dabosi, Y. Limouzin-Maire, On the mechanism for improved passivation by additions of tungsten to austenitic stainless steels, *Corrosion*, 39 (1983) 491-496.

[19] J.S. Kim, H.S. Kwon, Effect of tungsten on corrosion and kinetics of sigma phase formation of 25% chromium duplex stainless steels, *Corrosion*, 55 (1999) 512-521.

[20] K. Noda, T. Nishimura, H. Masuda, T. Kodama, Ion selective of rust formed on low-alloy steels under a cyclic wet-and-dry condition, *Journal of the Japan Institute of Metals*, 64 (2000) 767-770.

[21] I. Olefjord, B. Brox, U. Jelvestam, Surface composition of stainless steels during anodic dissolution and passivation studied by ESCA, *Journal of Electrochemical Society*, 132 (1985) 2854-2861.

[22] K. Hashimoto, K. Asami, An X-ray photo-electron spectroscopic study of the passivity of ferritic 19Cr stainless steels in 1 NHCl, *Corrosion Science*, 19 (1979) 251-260.

[23] M. Froment (Ed.), *Passivity of Metals and Semiconductors*, Elsevier, 1983.

Chapter 8

EFFECT OF TUNGSTEN ON HYDROGEN EMBRITTLEMENT

ABSTRACT

One of the significant challenges in the development of microalloyed steels is the so-called hydrogen embrittlement (HE) when these steels are applied in a hydrogen environment. Occurrence of HE of microalloyed steels may induce serious reduction in the ductility and load-bearing capacity, and cause cracking and catastrophic brittle failures at stresses below the yield stress due to the ingression of hydrogen into steels. Therefore, HE is of importance needs to be considered when applying a microalloyed steel in an environment that can be the sources of hydrogen due to the harmful effect of HE on microalloyed steels. In this chapter, the concept of HE is at first introduced. The factors affecting HE of steels are then described, followed by an introduction of strategies for the prevention of HE of steels. Then, reversible and irreversible hydrogen trapping sites in steels are classified based on the values of activation energy for hydrogen trapping. Next, typical existing investigations on the HE of microalloyed steels are briefly reviewed through introducing relevant experimental methods and main research findings. A cathodic electrochemical method is designed to do the hydrogen charging on circumferentially notched cylindrical specimens. HE susceptibility of microalloyed steels is assessed by slow strain rate test (SSRT), and the average hydrogen concentration in a hydrogen-charged specimen is evaluated by means of thermal desorption spectrometry (TDS) analysis. The effects of tungsten on the hydrogen desorption behaviour, activation energy for hydrogen desorption, notch tensile behaviour and fracture characterisation of microalloyed steels are then systematically analysed. Finally, the mechanism of hydrogen-induced embrittlement is discussed.

8.1. Concept of Hydrogen Embrittlement of Steels

Hydrogen embrittlement (HE), also known as hydrogen-assisted cracking, is a phenomenon that the ductility and load-bearing capacity are seriously reduced, and cracking and catastrophic brittle failures at stresses below the yield stress are caused due to the ingression of hydrogen into steels. There are four different categories of HE occurred in steels, which are distinguished either according to the source of hydrogen or the mechanism of damage:

- **Internal reversible HE**, so called because it takes place in the presence of solute hydrogen that can be removed by relevant treatment process such as baking heat treatment. It is a phenomenon that solute hydrogen diffuses to favoured sites where interatomic bonds are weakened, leading to internal crack initiation and propagation. Such HE occurs severely at near room temperature.
- **Internal irreversible HE**, whereby solute hydrogen presents in such high concentrations that hydrogen accumulates at various defects such as ferrite/cementite and inclusion/matrix interfaces with high trapping activation energy of hydrogen. The resulting high gas pressures may cause blistering or cracking.
- **Environmental HE**, which occurs when the steel is being subjected to external hydrogen gaseous or hydrogen sulphide. It differs from the internal HE because of its relation to the service environment. The failures occur after installation due to hydrogen being introduced from the service environment.
- **Hydrogen reaction embrittlement**, which is a phenomenon in which the hydrogen diffuses into the steel to precipitates or other defects and then generate gas bubbles - ‘blistering’. These reactions usually occur in environments containing hot hydrogen-bearing gases, resulting in the formation of blisters or expansions from which cracks may start to weaken the steel.

8.2. FACTORS AFFECTING HYDROGEN EMBRITTLEMENT OF STEELS

Several factors including characterisation of steel, environment and state of stress (internal or applied) affect to which extent embrittlement is induced by hydrogen. Once two of these three factors favour the occurrence of HE, an earlier failure is usually inevitable.

HE is usually associated with high strength steels. In general, as the strength of the steel goes up, so does its susceptibility to HE. High strength steels, such as quenched-and-tempered steels or precipitation-hardened steels, are particularly prone to HE. Other characteristics of steels including chemical compositions, microstructure and impurities may also affect the susceptibility of HE. High and ultrahigh strength martensitic steels, low alloy steels and maraging steels with yield strengths greater than 1400 MPa are extremely susceptible to HE, and even very low concentration of hydrogen can cause the occurrence of HE [1]. Lower strength martensitic and ferritic steels are less susceptible and require higher hydrogen concentrations for embrittlement to occur. Steels with high levels of impurities have a high susceptibility to HE.

The term "environment" mentioned here includes a wide variety of environments that can be the sources of hydrogen. Sources of hydrogen include heat treatment furnace atmosphere, breakdown of organic lubricants, steelmaking process (e.g., electric arc melting of damp scrap), fabrication of steel, processing steel parts working environment, welding, dissociation of high pressure hydrogen gas and even grinding, etc. In addition, operation of electrochemical surface treatments of steels such as etching, pickling, phosphate coating, corrosion removal, paint stripping and electroplating may also produce a considerable amount of hydrogen. HE can occur whenever atomic or protonic hydrogen is generated from a reaction in the hydrogen-containing or potential hydrogen generating environment. For example, steel can react with hydrochloric acid during pickling operation to diffuse hydrogen into the steel. Platers must be especially aware of HE effects since many pre-plating and plating operations can be potential sources of absorbable hydrogen. Since corrosion reactions are also generators of hydrogen, care in choosing the proper coating to prevent corrosion is also quite important. In general, any process producing atomic hydrogen at a steel surface will induce considerable hydrogen absorption in that steel. Therefore, steels should be evaluated carefully before they are used in the hydrogen-containing or hydrogen-generating environment. A steel should not be used unless data are available to prove that it is suitable for the planed service environment.

The embrittlement is strongly connected with locally high hydrogen concentrations which can be caused by stress-enhanced diffusion rates to lattice defects and reaction sites to initiate cracks. When the concentration of hydrogen exceeds a critical value in the defects, rapid brittle fracture occurs. Crack growth stops when the crack has grown through the hydrogen-enriched region or when the stress factor has decreased sufficiently. Both internal residual stress generated during processing and external stress applied affect significantly the occurrence of HE. Once a critical hydrogen concentration is reached and either the internal or external stress is sufficient, HE will take place. Failure induced by hydrogen is favoured or caused by applied stress, even static or of low intensity, and is often referred to as "delayed" or "retarded", since it may not take place immediately after tensile stress is applied.

Due to the harmful effect of HE on high strength steels, an insight in the interaction between HE affecting factors and failure behaviour is of crucial importance. Extensive investigations have been performed to understand the causes of HE occurring in various high strength steels including twining-induced plasticity (TWIP) steels [2], transformation induced plasticity (TRIP) steels [3], dual phase (DP) steels [4], high Mn steels [5], tempered martensitic steels [6], ferrite-martensite steels [7], stainless steels [8], maraging steels [8], ODS RAF steels [9] and microalloyed steels [10].

8.3. PREVENTION OF HYDROGEN EMBRITTLEMENT OF STEELS

Strategies for the prevention of HE should be developed based on the factors mentioned in Section 8.2 that may cause HE of steels. Basically, HE can be avoided by (i) preventing hydrogen absorption wherever possible, (ii) eliminating residual stress in the steel part before processing, and (iii) heat treating (e.g., baking) the steel part after processing to remove absorbed hydrogen before it can damage the part when use.

For practical purposes, the following methods can be used to prevent or minimise the occurrence of HE of steels:

1. Use of low strength (hardness) steels in case they meet the application requirements.
2. Use of steels with low levels of impurities.
3. Use of steels with low HE susceptible microstructure.
4. Reduction of residual and applied stresses.
5. Avoidance of exposure to hydrogen whenever possible.
6. Modification of the environment that may be the source of hydrogen to reduce hydrogen absorption.
7. Avoidance of cathodic cleaning, picking or activation treatment whenever possible, by use of alkaline soak cleaning and anodic cleaning.
8. Use of vapour degreasing or solvent cleaning to remove the grease, oil, or other contaminations before cleaning in aqueous solution.
9. Use of mechanical means, e.g., tumbling, grit blasting and vapour blasting, to remove oxide scale rather than pickling.
10. Use of surface plating and effective inhibitors to avoid or minimise HE.
11. Use of inhibited pickling and plating solutions that may cut down on the amount of steel dissolved and thereby reduce the amount of hydrogen generated.
12. Use of low embrittling electroplating processes that may cause either a lower pickup of hydrogen or in a deposit that allows easier removal of the absorbed hydrogen during the baking treatment.
13. Use of mechanical agitation in pickling and plating to quickly remove hydrogen generated.
14. Baking of the parts that are pickled and plated.
15. HE testing of the final parts using the methods described in ASTM standards, such as ASTM F519 [11], or other proven equivalent approaches to determine if the parts are susceptible to HE.

8.4. Hydrogen Trapping in Steels

HE of steels correlates closely with the attractive interactions between hydrogen atoms and structural imperfections in crystals, i.e., the hydrogen trapping phenomena at various structural imperfections in steels affect the susceptibility of HE [12]. The presence of hydrogen in solution in steels is mainly related to the small diameter of hydrogen atom and its capacity to diffuse and/or trap under a certain condition in solid state. Different factors such as environment, temperature, chemical composition, stress status and microstructural constituents can affect the hydrogen diffusion and trapping behaviour in steels. Almost all kinds of defects in steels including vacancies, dislocations, interfaces, micro-voids and grain boundaries can act as hydrogen trapping sites. The trapped hydrogen may diffuse to the stress-concentrated zone and microcracks that occurred at the tips of defects during loading, and an early failure is expected as the loading proceeds. There are generally two kinds of hydrogen trapping sites in iron and steels based on the trapping activation energy that is needed to escape from the trapping site to the normal lattice site. Table 8.1 presents the values of trapping activation energies of hydrogen for a number of trapping sites in iron and steels [13]. It can be seen that the trapping activation energies of hydrogen for defects like iron lattice, grain boundary, austenite/martensite interface, dislocation, austenite/dislocation boundary, microvoid and iron oxide interface are low, while those for ferrite/cementite interface and various nonmetallic inclusions such as chromium carbide, Y_2O_3, MnS, Al_2O_3, Fe_3C and a precipitate, TiC are high. The ferrite/cementite interface and interfaces of nonmetallic inclusions with high trapping activation energy of hydrogen are usually called irreversible trapping sites, whereas those with lower energies are termed as reversible or diffusible trapping sites.

Table 8.1. Trapping activation energies of hydrogen for a number of trapping sites in iron and steels [13]

Trapping site	Trapping activation energy (kJ/mol)	Reference
Reversible trapping sites:		
Iron lattice	5.4-7.1	[14]
Grain boundary	17.2-18.6	[15,16]
Austenite/martensite interface	22	[5]
Dislocation	26.4-26.8	[15,17]
Austenite/dislocation boundary	37	[5]
Microvoid	35.2-40	[15,18]
Fe oxide interface	47	[9,19]
Irreversible trapping sites:		
Ferrite/cementite interface	66.3-68.4	[6,20]
Cr carbide interface	67	[21]
Y_2O_3 interface	70	[19]
MnS interface	72	[22,23]
Al_2O_3 interface	79-86.2	[22,24]
Fe_3C interface	84	[19]
TiC interface	68-137	[25,26]

8.5. Investigations on the Hydrogen Embrittlement of Microalloyed Steels

Typical existing investigations, including experimental methods and main research findings, on the HE of microalloyed steels are summarised as follows:

- Michler and Naumann [27] performed heat treatments on some microalloyed steels with different chemical compositions in order to obtain a variety of microstructural combinations of ferrite/pearlite, bainite and martensite. Tensile tests on smooth specimens of these steels were then conducted in a gaseous hydrogen environment at 10 MPa and 20°C with a cross head speed of 0.1 mm/min (strain rate is 5.5×10^{-5} s^{-1}). They found that hydrogen assisted crack propagation, and hydrogen enhanced localised plasticity was the primary failure mechanism.
- For evaluating the hydrogen-induced cracking (HIC) susceptibility of submerged arc welded API 5L-X70 HSLA pipeline steels, Beidokhti et al. [28] conducted HIC resistance tests in a H_2S-saturated solution, consisting of 0.5% acetic acid and 5% sodium chloride dissolved in distilled water, at 25°C for 96 h according to the NACE Standard TM0284-96 [29]. They found that increased acicular ferrite content in the microstructure (more than 60%) improved the resistance of HIC. The welds containing more than 30% bainite and martensite/austenite in the microstructure failed significantly after completion of HIC. The precipitated titanium carbonitrides could act as beneficial hydrogen traps and delay cracking in hydrogen sulphide environment.
- Kim et al. [30] studied the effects of environmental and metallurgical factors on HIC of HSLA steels by performing HIC tests in nine different test solutions with varying combination of P_{H_2S} (the percent concentration of H_2S gas being contained in N_2 gas) and pH values for 96 h according to the NACE Standard TM0284-02 [31]. The results indicated that P_{H_2S} had a much bigger effect on HIC as compared to pH value, and diffusible hydrogen amount trapped reversibly in steels was more dominant in the initiation of HIC relative to hydrogen fluxed permeating through steels. HIC in microalloyed steels primarily nucleated at inclusion and/or clusters containing the aluminium and calcium oxides. They claimed that the HIC resistance

was determined by the diffusible hydrogen amount in HSLA steels with different microstructures when the steels had the same inclusion level.

- Kim et al. [32] conducted a research on the effect of post-weld heat treatment on HIC behaviour of high-strength process pipe steel in a sour environment according to the NACE Standard TM0284-03 [33]. They concluded that in acicular ferrite, Fe_3C particles acting as reversible trapping sites for hydrogen atoms were newly precipitated along the grain boundary of microalloyed steel by the post-weld heat treatment at 620°C. This led to an increase in diffusible hydrogen content in the steel that made it more susceptible to hydrogen-induced cracking.
- Nanninga et al. [34] studied three microalloyed pipeline steels (X52, X65 and X100) by tensile tests in a high-pressure (13.8 MPa) hydrogen gas environment. The results showed that the ductility of these steels was significantly degraded by gaseous hydrogen. HE increased with steel strength and increased the increasing the gas pressure. Hydrogen absorption occurred prior to reaching the yield point for these steels, but after yielding, the rate of absorption and diffusion became much more rapid due to dislocation assisted mobility. They concluded that this initial hydrogen probably led to the formation of the hydrogen-induced surface cracks, and the tensile specimens tested in hydrogen failed through a mechanism of surface crack initiation and growth that occurred during the necking of the tested specimens. The micro-mechanism of hydrogen cracking was quasi-cleavage fracture of ferrite grains.
- Alp et al. [35] tested the HE behaviour of a dual-phase microalloyed steel containing niobium, vanadium and titanium. They found that the susceptibility to HE in microalloyed steel was closely related to the microstructural state. Hydrogenated specimens with martensite islands in a ferrite matrix basically exhibited quasi-cleavage fracture with some ductile dimpling. The mode of fracture in charged specimens quenched from higher intercritical annealing temperatures was predominantly intergranular fracture along prior austenite grain boundaries and cracking of martensite laths.
- Opiela [36] studied the HE of welded joints for the XABO 960 microalloyed steel heavy plates by using tensile tests with an advance speed of 5 min/min. The flat samples for tensile tests (gauge length of 30 mm, and cross sectional area of 3 mm × 15 mm) were charged

with hydrogen electrolytically separated from 1 N solution of sulphuric acid. The results revealed that hydrogen accumulated on fronts of non-metallic inclusions, microcracks and on other crystal defects, where powerful exothermal reaction of recombination of hydrogen atoms into hydrogen molecules took place, leading to HE of welded joints. The HE of welded joints from XABO 960 microalloyed steel plates was evidenced by a distinct decrease in ductility and a slight decrease in strength. It was found that the welded joints were susceptible to hydrogen cracking in both the heat affected and the fusion zones.

- For studying the HE behaviour of microalloyed rail steels, Moon et al. [37] conducted tensile tests (strain rate of 0.1 s^{-1}) on smooth specimens which were hydrogen-charged by cathodic polarisation in an electrochemical cell containing 0.5 mol/l H_2SO_4 solution with a current density of 0.1 A/cm^2. They concluded that the rail steels were susceptible to HE as confirmed by the reduction of ductility. The degree of HE was higher in carbon-manganese rail steel as compared to the copper-molybdenum and chromium-copper-nickel rail steels.
- Banerjee and Chatterjee [38] studied the HE behaviour of a copper precipitation strengthened and niobium microalloyed HSLA 80 steel under cathodic charging conditions. They performed tensile tests on smooth specimens (gauge length of 36 mm and diameter of 4.2 mm) by using a slow strain rate technique (strain rates varied between 1.5×10^{-6} and 8.3×10^{-6}) in synthetic sea water containing 0.01 kg KCl, 0.045 kg KBr, 0.055 kg $MgCl_2$, and 0.110 kg $CaCl_2$ in 1 litter of distilled water. The results showed that the steel is prone to HE in the as-received quenched and tempered state at an electrode potential of -800 mV (SCE) and beyond. Hydrogen induced microvoid formation and increasing quasi-cleavage tendencies confined at the peripherial regions with decreasing the electrode potential.
- Vehovar [39] carried out an investigation on the resistance of two microalloyed structural steels to HE due to the action of trapped internal atomic hydrogen by using a cathodic charging method at room temperature in deaerated 1 N H_2SO_4 containing 10 mg/l of As_2O_3. After cathodic charging for different period of time, all the smooth specimens were taken out from the corrosion cell and finally tested by means of tensile tests. The results revealed that fine-grained structural microalloyed steels appeared to be more or less susceptible to hydrogen induced cracking due to various strong interactions

between the absorbed hydrogen and the various alloying elements or their carbides that acted as hydrogen traps.

- Ravi et al. [40] studied the effect of molybdenum on the resistance to HE of high sulphur microalloyed steels by slow strain rate tensile testing with a cross head speed of 0.1mm/min (strain rate of $0.67 \times 10^{-4}\ s^{-1}$) after exposing the tensile specimens to a 7 N H_2SO_4 + 1 g/l thiocarbamide solution for 120 h. They concluded that precipitates in microalloyed steels could act as deep hydrogen traps and the hydrogen absorbed by the steel could only be recovered above 500°C. The addition of molybdenum in the steels led to improvement in the HE resistance, and the optimal molybdenum content was around 0.3%. Further addition of copper, chromium and nickel to a molybdenum bearing steel resulted in decreased hydrogen content in the steel and thereby further improved the HE resistance.

8.6. Testing Procedure for Assessing Hydrogen Embrittlement

Although HE of microalloyed steels has been investigated extensively, there is still no relevant publication available that addresses HE behaviour in tungsten-containing microalloyed steels so far, let alone a systematic investigation on the quantitative relationships among hydrogen concentration, tungsten content and HE behaviour. As an alloying element with growing scientific and industrial interests, tungsten has attracted an increasing attention due to its positive effect in microalloyed steels. It is therefore worthwhile to conduct an in-depth research with the purpose of understanding the effects of tungsten on the HE behaviour of microalloyed steels. In this section, the testing procedure for assessing HE will be introduced. And then in the following sections, the effects of tungsten on the HE behaviour of several microalloyed steels will be systematically analysed, and the mechanism of hydrogen-induced embrittlement in these steels with different tungsten contents will be discussed.

8.6.1. Preparation of Notched Specimens

Circumferentially notched cylindrical specimens with a notch section diameter of $\Phi = 6$ mm, a notch angle of $\theta = 60°$ and a notch root radius of $r = 0.1$ mm were machined for slow strain rate test (SSRT). The detailed dimensions and the appearance of a notched specimen are shown in Figure 8.1 [13]. The notched specimens were prepared by a computerised numerical control (CNC) machine. The stress concentration factor was 4.9. Local stress concentration will be caused at the notch during tensile deformation. Several studies have indicated that notch accelerates the diffusion of hydrogen from the other zone to the notch root, and therefore helps to increase the hydrogen mobility in the specimens during loading [41,42].

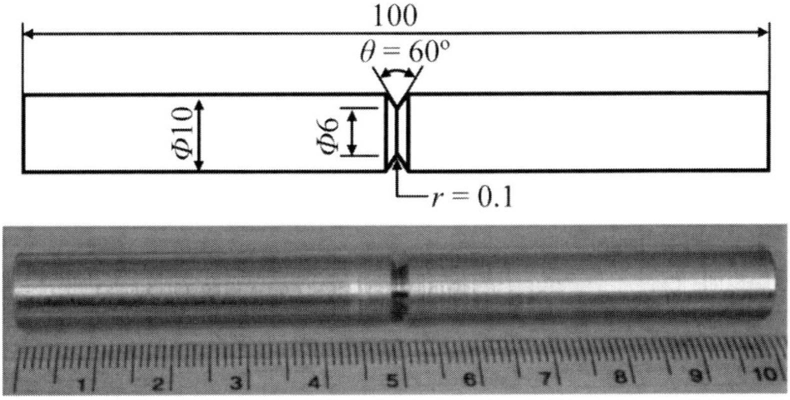

Figure 8.1. Dimensions (in mm) and appearance of a notched specimen for the SSRT [13].

8.6.2. Electrochemical Hydrogen Charging

Hydrogen was introduced into the notched specimens by cathodic electrochemical charging in a solution of "3 wt.% NaCl + 0.3 wt.% NH_4SCN" under the current densities of 1 to 20 A/m^2 for 48 h at room temperature. The roles of Na^+ and Cl^- ions in the solution are electrolytes and the acid hydrolysis of NH^+ supplies H^+ ions, which remain as adsorbed atoms on the specimen surface after cathodic reduction. The recombination of H^+ ions into a gaseous form is inhibited by the CNS^- ions [43]. The cathodic electrochemical hydrogen charging is schematically illustrated in Figure 8.2. Due to the sensitivity of hydrogen charging to the surface roughness, all the surfaces of the notched specimens were mechanically polished with #1200 SiC sand papers before hydrogen charging processes.

Immediately after hydrogen charging, the specimens were mechanically polished with #1200 SiC sand papers and then electroplated with a cadmium coating to prevent hydrogen desorption. Cadmium-coated layer is also an effective barrier to hydrogen escape from the specimens during the subsequent SSRTs. For cadmium coating, specimens were connected to the cathode of the power, and then they were inserted into a mixed electrolyte solution of "$Cd(BF_4)_2$ + NH_4BF_4 + H_3BO_4 + water". Cadmium metal block put in this electrolyte solution was connected to the anode of the power. Coating time was controlled to be 5 min for a hydrogen-charged specimen under the current of 0.5 A. The experimental apparatus for cadmium coating is shown in Figure 8.3. Following cadmium coating, the specimens that were hydrogen charged for 48 h were kept at ambient for 24 h to homogenise the hydrogen distribution. Previous experimental study has indicated that a total time of 72 h for charging and homogenising is effective and reasonable for obtaining uniformly-distributed hydrogen in the specimens [44]. Then, the specimens were subjected to SSRTs.

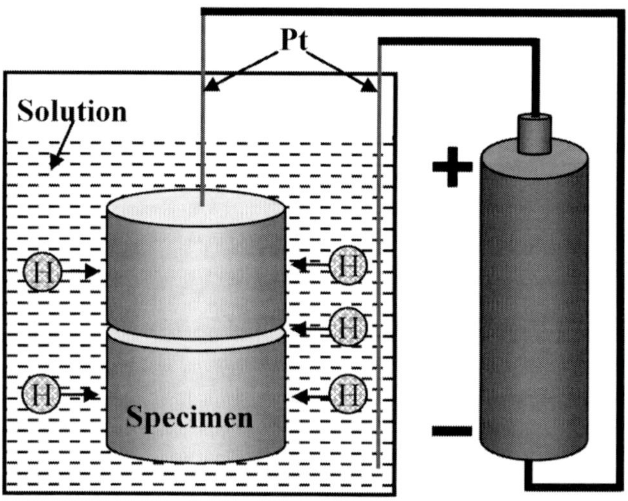

Figure 8.2. Schematic illustration of the cathodic electrochemical hydrogen charging.

Figure 8.3. Experimental apparatus for cadmium coating.

8.6.3. SSRT and Thermal Desorption Spectrometry (TDS) Analysis

SSRT was performed on an INSTRON 8501 tensile testing machine at a constant stroke speed of 0.005 mm/min (strain rate is 8.3×10^{-7} s^{-1}). The cadmium-coated layer was electrochemically removed from the specimen surface immediately after SSRT. The process for removing cadmium from the specimen surface is similar to that of cadmium coating. The only difference is that specimen should be connected to the anode, and cadmium metal block should be connected to the cathode of the power during the removal process of the cadmium-coated layer. The fracture surfaces of the tested specimens were observed on SEM using an accelerating voltage of 15 kV to determine the failure mode.

The average hydrogen concentration in a hydrogen-charged specimen was evaluated by means of TDS analysis using a gas chromatography (Agilent GC 7890A) at a heating rate of 100 K/h. The temperatures were raised from ambient temperature to 773 K to obtain different hydrogen desorption profiles. During TDS analysis, the emitted gas was collected at an interval of 5 min under a continuous helium gas flow.

The total hydrogen that was desorbed below 600 K during heating was regarded as the diffusible hydrogen because it can diffuse out of the specimens if the specimens are exposed at room temperature. In addition, this hydrogen is essential for the occurrence of hydrogen-induced embrittlement, although it would be either lattice hydrogen or weakly trapped hydrogen [20]. The hydrogen trapping sites were identified based on the activation energy of hydrogen desorption. Other heating rates of 200 K/h and 300 K/h were also employed for determining the activation energy of hydrogen desorption.

8.7. Effect of Tungsten on Hydrogen Desorption Behaviour

Figure 8.4 shows the typical hydrogen desorption curves of the 0W steel [13]. As shown in Figure 8.4a, no hydrogen desorption curve appears during heating for the uncharged specimen. The hydrogen-charged specimens, however, exhibit additional hydrogen desorption peaks at a temperature of around 360 K, which are undoubtedly caused by hydrogen charging. The magnitude of the peak desorption rate indicates the amount of charged hydrogen in the specimen, and a higher peak corresponds a higher hydrogen concentration. Under the charging current densities of 5 and 10 A/m^2, for example, the measured hydrogen concentrations are 0.55 and 0.98 mass ppm, respectively, by means of TDS analysis. Figure 8.4b illustrates the effects of heating rate on the hydrogen desorption curves. It is clear that the higher the heating rate during TDS analysis, the higher the temperature to the peak desorption rate is expected.

Figure 8.5 shows the effects of tungsten on the profile of hydrogen desorption curves [13]. It can be seen that the peak values decrease gradually with the increase of tungsten, and the peak shifts to a higher temperature side as tungsten is added. Tungsten addition induces a reduced peak area (an area underneath the hydrogen desorption curve). The peak area is proportional to the hydrogen concentration that charged in a specimen. Under the charging current density of 5 A/m^2, the measured hydrogen contents are 0.55, 0.52, 0.48 and 0.42 mass ppm, for the 0W, 0.1W, 0.5W and 1W steels, respectively. Therefore, tungsten addition decreases the amount of diffusible hydrogen in microalloyed steels under the same hydrogen charging current density.

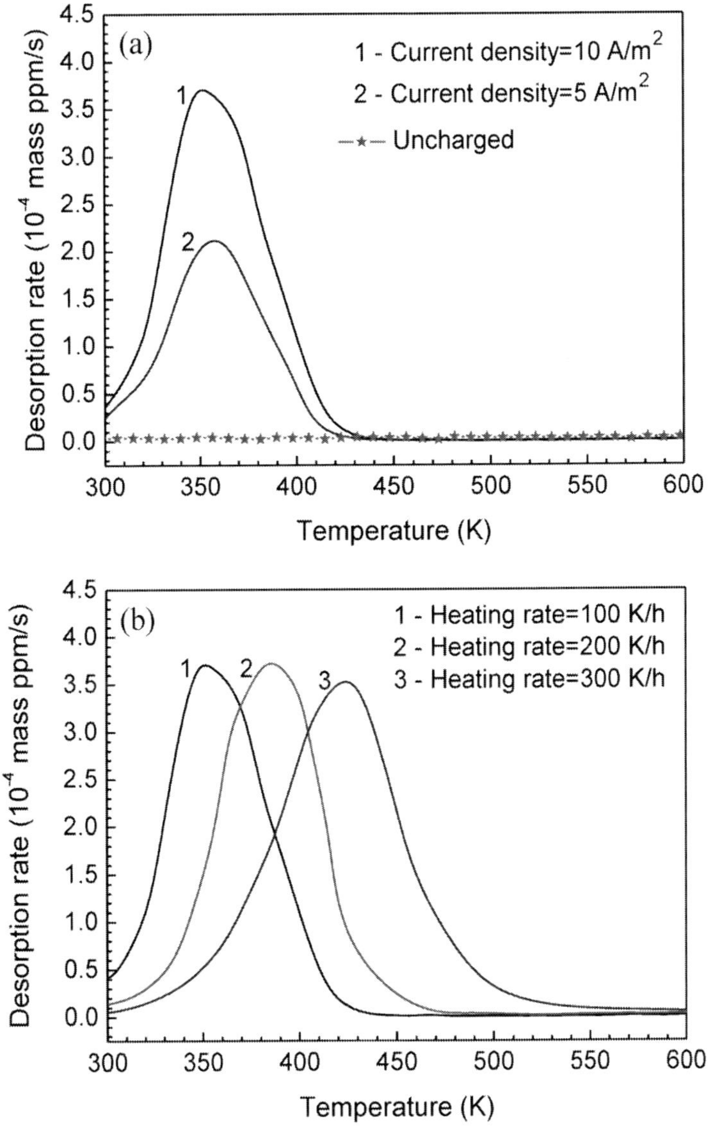

Figure 8.4. Hydrogen desorption curves of the 0W steel with (a) heating rate of 100 K/h and (b) current density of 10 A/m^2 [13].

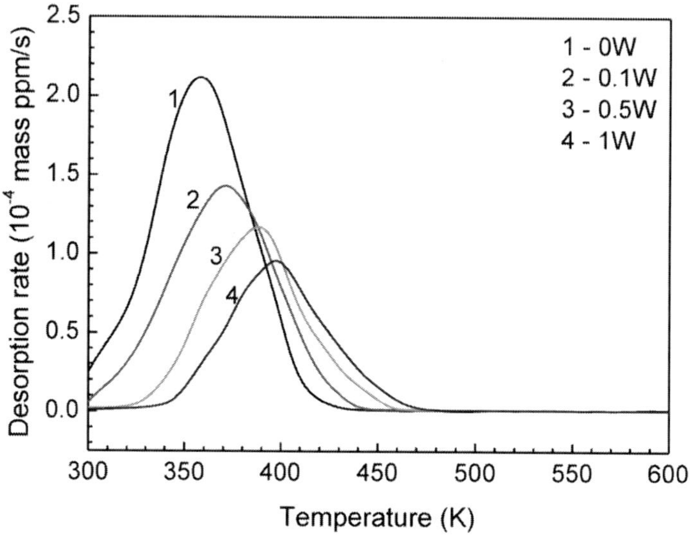

Figure 8.5. Effects of tungsten on the profile of hydrogen desorption curves under the hydrogen charging current density of 5 A/m^2 and heating rate of 100 K/h [13].

The dependence of measured hydrogen concentration on charging current density of the tested steels is illustrated in Figure 8.6 [13]. It is clear that the hydrogen concentration increases with charging current density for all the specimens. Under a given charging current density, the hydrogen concentration shows a decreasing trend with the increase of tungsten, which is consistent with the result obtained from Figure 8.5. It is therefore more difficult for hydrogen to permeate in tungsten-added steels during hydrogen charging as compared to the 0W steel.

The permeability of hydrogen in steels is affected by the microstructural features including phase constituents, grain boundaries, grain size and shape, vacancies and dislocations, interfaces with nonmetallic inclusions, precipitate particles and voids. The work of Wang et al. [45] indicated that the main diffusion paths for hydrogen were the grain boundaries of equiaxed ferrite and ferrite/carbide interfaces of pearlite in the ferrite/pearlite steel, ferrite/carbide interfaces of long needle-like or lath-like bainite in bainitic steel, and boundaries of the blocky grain boundary ferrite and interfaces in the steel with AF structure. They concluded that the permeability and diffusivity of hydrogen in steels followed a decreasing order of equiaxed ferrite and pearlite, AF, and bainite.

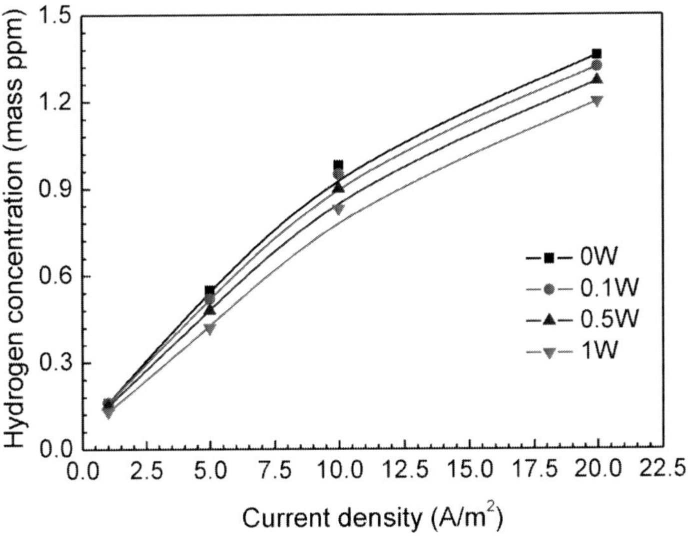

Figure 8.6. Dependence of hydrogen concentration on charging current density [13].

The 0W steel consists mainly of polygonal ferrite and scattered pearlite (see Section 4.3), indicating the highest permeability and diffusivity compared to the tungsten-containing steels. This is attributed to the ferrite/cementite interface in scattered pearlite and the grain boundaries of ferrite provide effective paths for hydrogen diffusion because a large amount of polygonal grains contribute to a large volume fraction of grain boundaries available as a hydrogen high diffusivity path [46]. As tungsten is added, the amount of AF is increased. AF, which is characterised by short ferrite needles with a basket-weave feature, has randomly-oriented grain boundaries and high dislocation density due to tangled dislocations in it [47]. Hydrogen can be trapped reversibly at the tangled dislocations and then induces decrease in permeability and diffusivity of hydrogen in the steel. As described in Section 4.4, the interlamellar spacing of pearlite decreases significantly after tungsten addition. For the tungsten-containing steels, the relatively low permeability and diffusivity of hydrogen might also be contributed to the tungsten-induced low interlamellar spacing of pearlite. The higher the amount of AF as the tungsten content is increased, the lower the permeability and diffusivity of hydrogen are then expected. After 1% tungsten is added, the microstructure consists of mixed AF and bainite. The lowest permeability and diffusivity of hydrogen in the 1W steel should be attributed to the high dislocation density of bainite and the cementite that is precipitated between the ferrite fingers acting as an obstacle for hydrogen transport [45, 46]. As a result, the pre-charged hydrogen concentration presents a decreasing trend with tungsten under the same hydrogen charging current density, as shown in Figures 8.5 and 8.6.

8.8. Effect of Tungsten on Hydrogen Desorption Activation Energy

Hydrogen in steel is prone to be trapped at various structural defects such as dislocations and grain boundaries. To identify the hydrogen trapping sites, the activation energy of hydrogen desorption is calculated by the Kissinger equation as [48]:

$$\frac{\partial \ln(\phi / T_p^2)}{\partial (1/T_p)} = -\frac{E_a}{R} \tag{8.1}$$

where ϕ is the heating rate in K/s, T_p is the peak temperature in K, E_a is the activation energy of hydrogen desorption in kJ/mol, and R is the gas constant. After performing TDS analysis using different heating rates (100, 200 and 300 K/h in the current work) and the determination of the corresponding peak temperatures for a trap in the hydrogen desorption curves, the slope of $\ln(\phi / T_p^2)$ versus $(1/T_p)$ plot allows to obtain the value of E_a that corresponds to a specific hydrogen trapping site. The activation energies of hydrogen desorption are calculated to be 12.75, 16.19, 21.57 and 25.82 kJ/mol for the 0W, 0.1W, 0.5W and 1W steels, respectively, based on Eq. (8.1). As analysed in Section 5.6, the dislocation density increases gradually with an increase of the tungsten content from 0 to 1% in the microalloyed steels.

According to Table 8.1, the calculated E_a in both the 0W and 0.1W steels should correspond to the hydrogen in lattice and trapped by reversible trapping sites such as grain boundaries and dislocations [20]. The measured diffusible hydrogen in the 0.5W steel is probably trapped by grain boundaries and dislocations since the activation energy and peak temperature are very similar to the reported ones [44,49]. The E_a value for the 1W steel, 25.82 kJ/mol, makes us ensure that the hydrogen may be trapped in dislocations, which is consistent with the result ($E_a \approx 26.8$ kJ/mol) obtained by Choo and Lee [15].

8.9. EFFECT OF TUNGSTEN ON NOTCH TENSILE BEHAVIOUR

As indicated in Figure 8.6, the hydrogen concentrations are different in the steels with different tungsten additions even though under the same hydrogen charging current density. In this chapter, the notch tensile behaviour of the steels with different hydrogen concentrations will be evaluated only under different hydrogen charging current densities, while the corresponding real hydrogen concentrations can be obtained directly from Figure 8.6.

Figure 8.7 shows the tensile stress versus displacement curves of the notched specimens that were charged under different hydrogen charging current densities [13]. In this figure, A0 indicates the specimen that was uncharged, and A1, A5, A10 and A20 represent the specimens that were charged under current densities of 1, 5, 10 and 20 A/m^2, respectively. For the 0W steel, all the specimens show yielding and necking behaviour before final fracture, and the ultimate notch stress and the displacement to fracture (D_f) decrease gradually with the increase of current density (Figure 8.7a). After 0.1% and 0.5% tungsten additions, the specimens without hydrogen charging exhibit a similar notch tensile behaviour to that of the 0W steel. Differently, the notch stress shows an increasing trend with the increase of tungsten content, and the displacement to fracture is reduced after tungsten addition, as shown in Figure 8.7b and c. The hydrogen-charged specimens exhibit an early fracture before obvious yielding, and the higher the current density, the earlier the fracture occurs. The addition of 1% tungsten has significantly changed the notch tensile behaviour. As shown in Figure 8.7d, fracture occurs before an ultimate notch stress is obtained for all the tested specimens. When the current density reaches 10 A/m^2 or a much higher value, the curve of the 1W steel appears to be a straight line, which means no obvious yielding and necking behaviour is observed before the final fracture occurs (Figure 8.7d). In addition, there is no significant change in the slope of notch tensile curve with varying tungsten contents and hydrogen charging current densities, indicating both tungsten and hydrogen have no significant effect on the elastic deformation before the final fracture occurs.

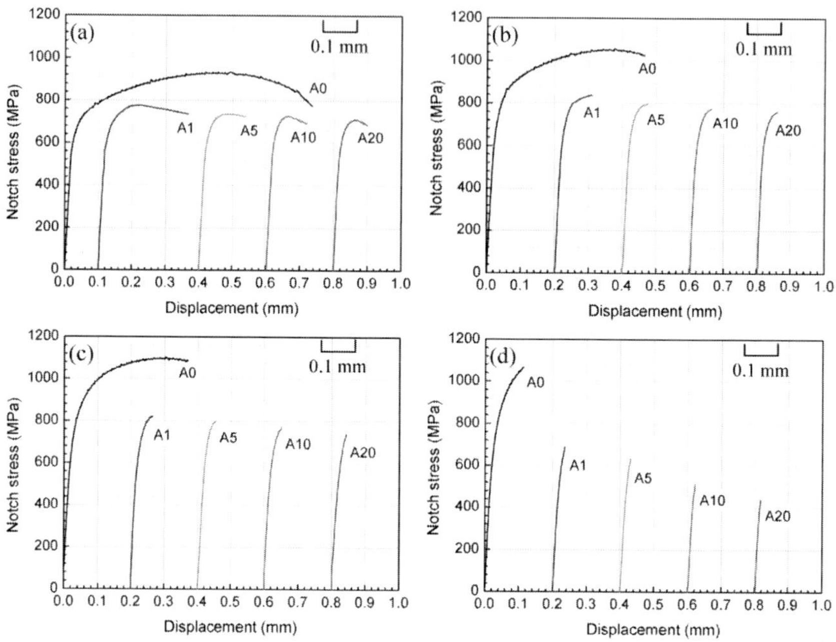

Figure 8.7. Notch tensile curves of the (a) 0W, (b) 0.1W, (c) 0.5W and (d) 1W steels charged under different current densities [13].

Figure 8.8 shows the variations of fracture stress (σ_f) and D_f with tungsten content under different hydrogen charging current densities [13]. As shown in Figure 8.8a, the σ_f of the uncharged specimens increases gradually at first, then to a maximum value at the tungsten content of 0.5%, and finally it drops slightly when the tungsten content is higher than 0.5%. The values of σ_f are reduced from 774.6 to 735.0, 1028.2 to 837.1, 1086.8 to 820.8, and 1065.2 to 688.8 MPa for the 0W, 0.1W, 0.5W and 1W steels, respectively, after hydrogen charging under current density of 1 A/m^2. The drop of σ_f shows an increasing trend with tungsten, as indicated by the arrows in Figure 8.8a. Among the hydrogen-charged specimens, the σ_f increases after 0.1% tungsten addition, and then it decreases slightly when the tungsten content reaches 0.5%. When the tungsten content is increased to 1%, a sharply decreased σ_f is observed.

As shown in Figure 8.8b, the D_f of the uncharged specimens shows a decreasing trend with tungsten. The arrows in Figure 8.8b indicate that hydrogen charging induces significant reduction of D_f, and the drop of D_f decreases gradually with tungsten. Moreover, the greater the current density, the smaller the values of σ_f and D_f are expected. Based on Figures 8.6 and 8.8, it can be concluded that both σ_f and D_f exhibit a decreasing trend with the increase of hydrogen concentration because the charged hydrogen concentration increases with current density.

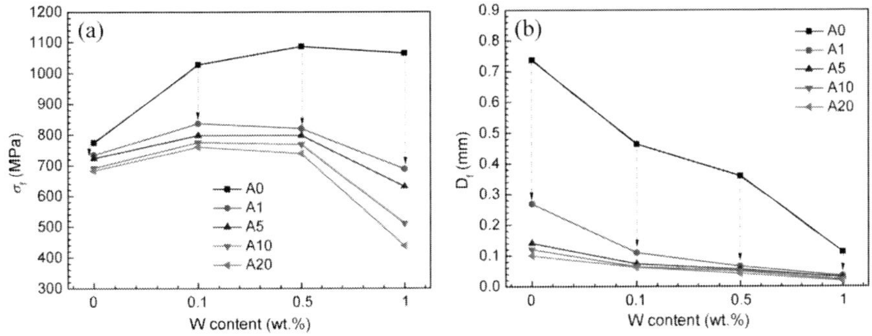

Figure 8.8. Dependences of (a) σ_f and (b) D_f on tungsten content [13].

Wang et al. [20] studied the tensile fracture behaviour of quenched-and-tempered AISI 4135 steel at 1450 MPa on both the smooth and notched specimens. They found that the notched specimens did not show general yielding before an abrupt fracture, and all the stress versus displacement curves were appeared as straight lines. In this chapter however, all the notched 0W, 0.1W and 0.5W specimens that are uncharged still present similar tensile behaviour to the smooth specimens which are all characterised by three distinct stages including elastic deformation, plastic deformation and fracture. The 1W specimen exhibits an early fracture before the ultimate notch stress is reached. One of the reasons should be the microstructural characteristics that contribute to lower tensile strength of the studies steels, as presented in Sections 4.3 and 5.5. Ferrite has a BCC crystal structure, in which there are 48 slip systems. Ferrite grains possess high ductility because dislocations can readily move and multiply to produce large permanent changes in shape of steel structures [50]. The 0W steel is mainly composed of PF and pearlite, and the dislocation density in this steel is very low relative to that in the tungsten-containing steels. A low dislocation density will cause a reduced amount of hydrogen trapped at dislocations, and the hydrogen will not accumulate rapidly due to the relatively high mobility of hydrogen during tensile deformation. Therefore, the hydrogen-charged 0W specimen still shows obvious three-stage deformation behaviour due to the high ductility of PF and low density of dislocation in the matrix even though the ductility is reduced due to the HE susceptibility. Several studies have indicated that hydrogen facilitates the emission and motion of dislocations [51,52]. As the mobility of dislocation is enhanced, a decreased strength will be obtained during tensile deformation. The work of Lü et al. [53] indicated that dislocation movement did not initiate when the studied steel was put into the hydrogen gas with a pressure of 0.1 MPa for 3 h. After remaining in the hydrogen gas for 6 h, however, hydrogen started to move as the charged hydrogen concentration was increased to a critical value. Therefore, hydrogen cannot facilitate the dislocation emission and motion until a critical hydrogen concentration is reached. The present work shows that the tensile strength of the 0W specimen decreases greatly even though only 0.16 mass ppm hydrogen (curve A1 in Figure 8.7a) is charged. This decrease is thought to be attributed to the high hydrogen concentration in the specimen, and the critical hydrogen concentration has been reached.

Tungsten addition induces significant change in the microstructural constituents, as described in Chapter 4. The precipitates are refined as the tungsten content is increased from 0 to 1%, as shown in Figures 3.8 and 3.9 of Chapter 3. At the same time, dislocation density is significantly enhanced, resulting increased strength and decreased ductility [54]. Dislocations can act as effective traps for hydrogen, thereby reducing the mobility of hydrogen atoms. Therefore, the higher dislocation density of the tungsten-added steels would be expected to allow slower diffusion of hydrogen during tensile tests. When a moving dislocation meets a precipitate, the dislocation most likely will lose the hydrogen proton to the irreversible trap. Several studies have indicated that fine precipitates are more effective for hydrogen trapping relative to the coarse ones [55,56]. Studies on precipitates have shown that the trapping capacity also depends on the lattice mismatch, i.e., whether the precipitated particles are coherent or incoherent [12,25]. The work of S.M. Lee and J.Y. Lee [12] showed that the trap activation energy of hydrogen increased with an increase of the precipitated incoherent particle size which is attributed to the changes of interfacial free energy. Therefore, the role of trapping upon hydrogen embrittlement is still not clear. Trapping might have a positive effect since the amount of diffusible hydrogen is strongly reduced. In the other hand however, trapping might be detrimental when hydrogen weakens the structural integrity, e.g., grain boundaries [57]. Hydrogen accumulation at locations of high triaxial stress such as precipitate/matrix interface, dislocations and grain boundaries may weaken Fe-Fe bonds [58]. It has been reported that HE increases with an increase of the strength levels of various steels including microalloyed steel [59], maraging steel [60] and chromium-molybdenum steel

[61]. These combined effects are expected to cause an early fracture as the tungsten content is increased, and the higher the pre-charged hydrogen concentration, the more susceptible of HE is expected under the same tungsten content, as shown in Figure 8.7.

8.10. Effect of Tungsten on Fracture Characterisation

Figure 8.9 shows the fracture surfaces of the notched 0W specimens with and without hydrogen charging [13]. At low magnification, the fracture characteristics of both the charged and uncharged specimens along the notch roots are observed to be somewhat different from that in other locations (Figure 8.9a and b). Observations on the uncharged specimen at a higher magnification show that a ductile fracture region (width is about 610 μm) which is composed of small dimples exists in the vicinity of the notch root (Figure 8.9c), while full cleavage fracture occurs in the central zone (Figure 8.9e). Also, the depth of dimples shows a decreasing trend along the direction from the margin of the notch root to the centre, and an immediate cleavage fracture is caused beyond the ductile fracture region from the notch root, which presumably corresponds to the crack initiation at the notch root and propagation towards to the central region till a final cleavage fracture is caused. Hydrogen charging has significantly changed the fracture mode of the tested specimen. As shown in Figure 8.9d, the hydrogen-charged specimen exhibits a brittle quasi-cleavage fracture in the vicinity of notch root. Comparing Figure 8.9c and d, the width and the location of the quasi-cleavage region are found to be same as that of the uncharged specimen, indicating a ductile to quasi-cleavage fracture transition of the 0W steel has taken place after hydrogen charging under current density of 1 A/m^2. In addition, hydrogen charging does not show noticeable effect on the fracture mode in the central zone, i.e., the hydrogen-charged specimen still exhibits a full cleavage fracture feature in the central zone which is characterised to be the same mode as the uncharged specimen, as compared in Figure 8.9e and f.

Figure 8.10 shows the fracture surfaces of the notched 0.1W and 0.5W specimens that were uncharged and hydrogen-charged under current density of 1 A/m^2 [13]. The fracture surfaces of the uncharged 0.1W and 0.5W specimens exhibit a similar characteristic to that of the 0W specimen, i.e., both the fracture surfaces are composed of a ductile fracture region near the notch root and a full cleavage fracture region in other locations. By comparing Figures 8.9c, 8.10a and 8.10c, it can be found that the ductile fracture region is reduced after tungsten addition, and the greater the tungsten content, the narrower the ductile fracture region is observed. The ductile fracture occurred in the vicinity of the notch root turns to be quasi-cleavage fracture after hydrogen charging, as indicated in Figure 8.10b and d. This finding is consistent with the result obtained from Figure 8.9 that hydrogen benefits the ductile to quasi-cleavage fracture transition and thus promotes the occurrence of embrittlement in microalloyed steels with and without tungsten additions.

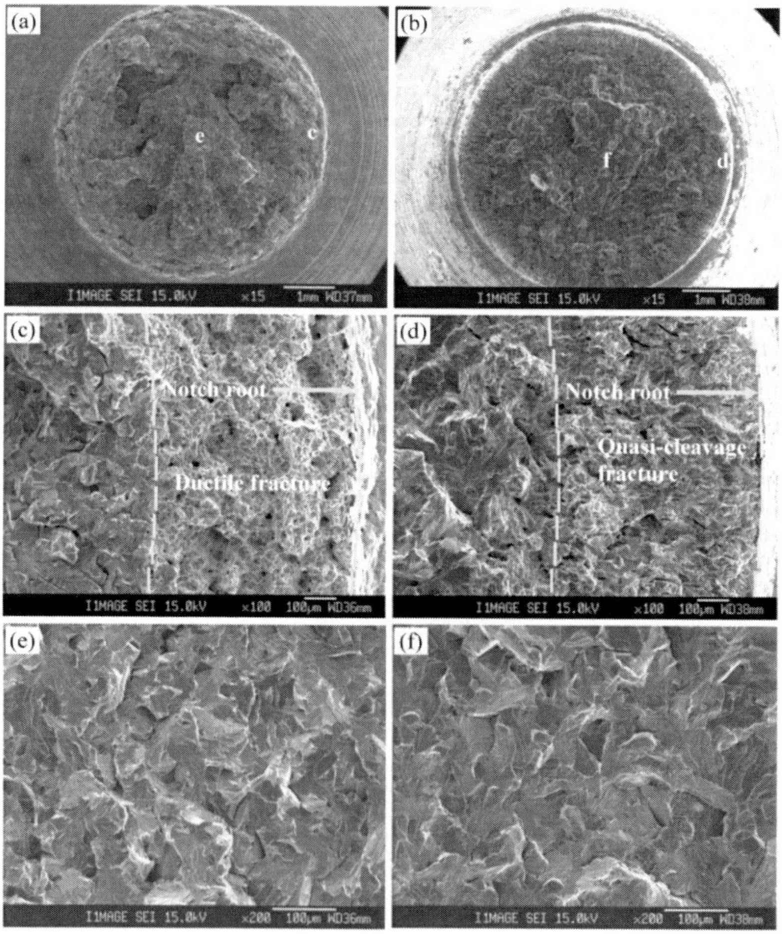

Figure 8.9. Fracture surfaces of the 0W specimens that were (a,c,e) uncharged and (b,d,f) hydrogen-charged under current density of 1 A/m^2 [13].

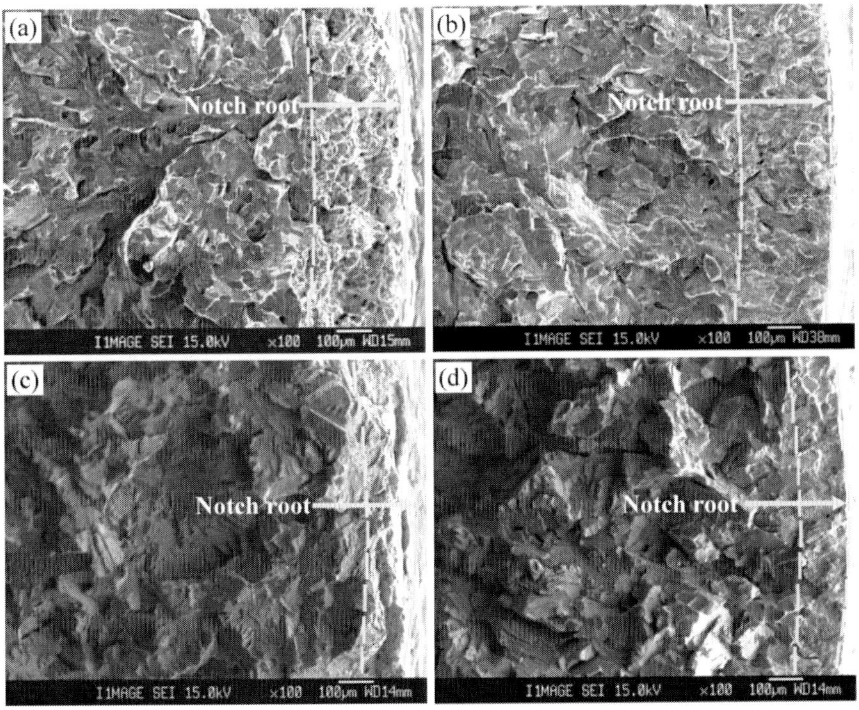

Figure 8.10. Fracture surfaces of the (a,b) 0.1W and (c,d) 0.5W specimens: (a,c) uncharged, and (b,d) hydrogen-charged under current density of 1 A/m^2 [13].

Figure 8.11 presents the fracture surfaces of the notched 1W specimens with and without hydrogen charging [13]. As shown in Figure 8.11a, no ductile fracture region is observed on the uncharged specimen, and a full cleavage fracture has been caused after 1% tungsten addition. Hydrogen does not show significant change in the fracture morphology, and the fracture feature of the hydrogen-charged specimen still shows a full cleavage mode, as shown in Figure 8.11b.

The fracture surfaces have been changed as the current density increases. As shown in Figure 8.12 [13], quasi-cleavage fracture still occurs in the vicinity of notch root of the 0W specimen after hydrogen charging under current density of 10 A/m^2. Differently, the width of the quasi-cleavage region has been decreased as the current density is increased from 1 to 10 A/m^2, as indicated in Figures 8.9d and 8.12.

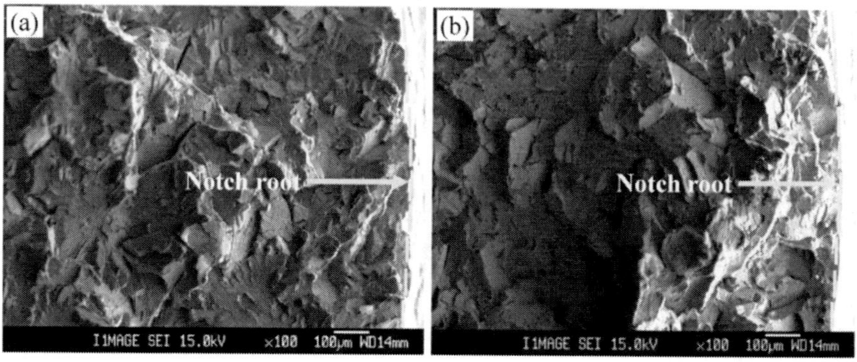

Figure 8.11. Fracture surfaces of the 1W specimens that were (a) uncharged and (b) hydrogen-charged under current density of 1 A/m^2 [13].

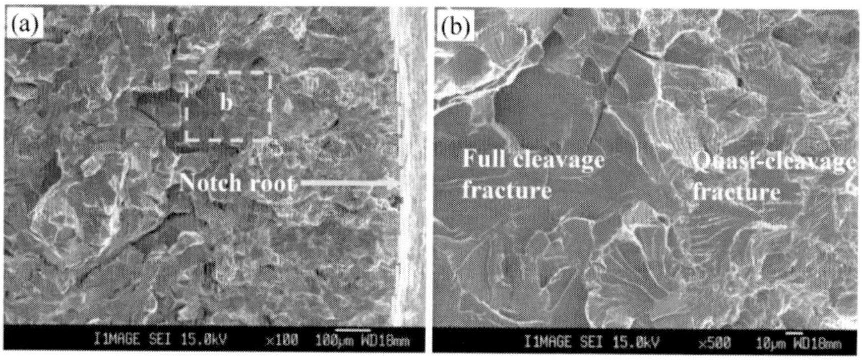

Figure 8.12. Fracture surfaces of the 0W specimen that is hydrogen-charged under current density of 10 A/m^2 [13].

Figure 8.13 shows the fracture surfaces of the tungsten-containing specimens that were hydrogen-charged under current density of 10 A/m^2 [13]. Two distinct fracture regions, i.e., a quasi-cleavage fracture region in the vicinity of the notch root and a full cleavage fracture region in other locations, are found on both the 0.1W and 0.5W specimens, as shown in Figure 8.13a through d. Small amounts of microcracks are observed at the edge of the notch root of the 0.1W specimen, as shown in Figure 8.13a and b. The characteristic and quantity of microcracks have been changed when the tungsten content increases.

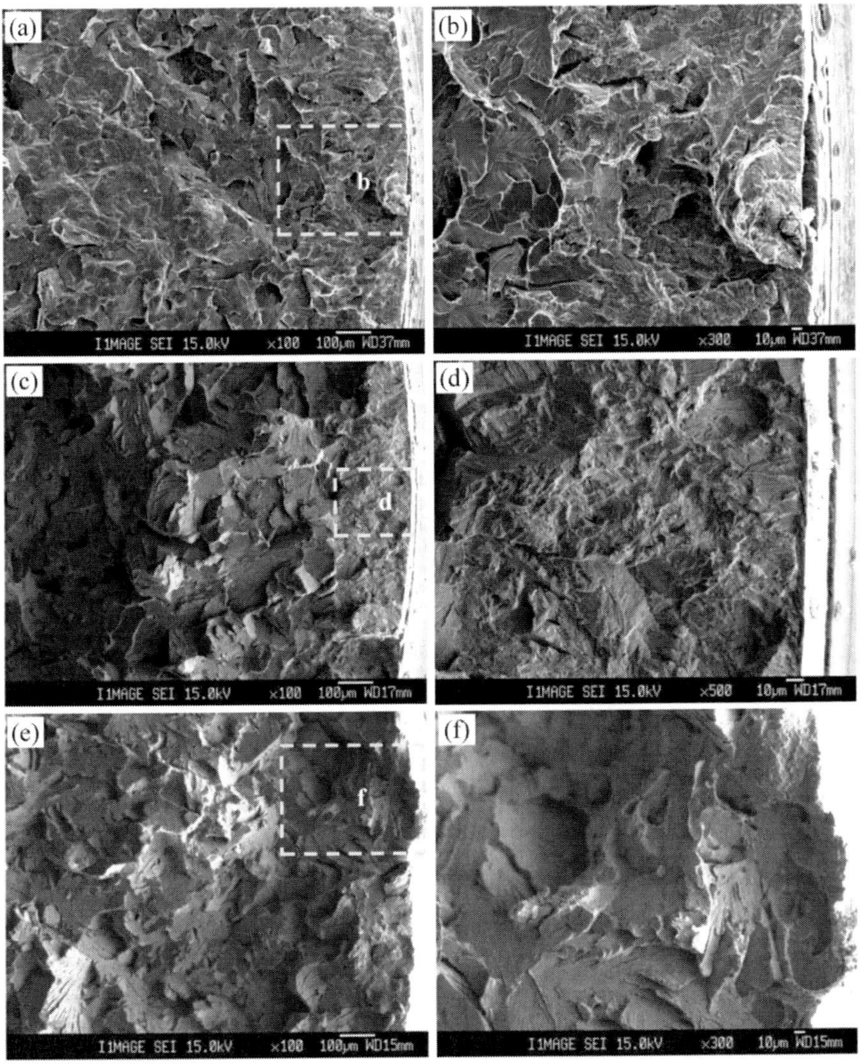

Figure 8.13. Fracture surfaces of the (a,b) 0.1W, (c,d) 0.5W and (e,f) 1W specimens that were hydrogen-charged under current density of 10 A/m^2 [13].

As indicated in Figure 8.13c and d, large numbers of through microcracks along the margin of the notch root are found as the tungsten content is increased from 0.1% to 0.5%. For the 1W specimen, however, the fracture surface shows a full cleavage mode, and no microcrack is found along the notch root, as shown in Figure 8.13e and f.

As analysed above, the width of ductile region in the vicinity of notch root shows a decreasing trend with tungsten, and no ductile fractured region is observed on the 1W specimen. The fracture surface of the notched specimen exhibits different characterisations from that of a smooth tensile specimen tested in air which is characterised by ductile dimples throughout the fractured cross section, as described in Section 5.5. One of the reasons should be the notch sensitivity of the tested material. Previous studies indicated that hydrogen was inclined to be accumulated at the notch root, and the hydrogen effect at the vicinity of notch root was much severer than that in a smooth specimen condition [41,44]. This phenomenon has been confirmed by Wang et al. [20] that fracture initiates in the vicinity of the notch root in a ductile mode for the notched specimens, and the area fraction of ductile region decreases with an increase of stress concentration factor. There is still a lot of debate on the hydrogen-induced degradation in steels, and several mechanisms have been proposed and discussed [62-64]. Among them, the decohesion theory and internal pressure theory have been adopted by many researchers in the investigation of HE in cast Fe-C steel [65], high-Mn steel [5], forged carbon steel [66], pipeline steel [67] and high strength steel [68] etc. with a varying hydrogen concentration from 0 to 26 mass ppm. In the decohesion theory, the presence of hydrogen atoms weakens atomic bonding force. The quasi-cleavage fracture occurred after hydrogen charging has been found to be attributed to decohesion of interface boundaries [5]. The internal pressure theory suggests that hydrogen atoms combine into hydrogen gas molecules at or near the interface of second phase particles (such as precipitates and cementite) and metallic matrix. The hydrogen molecules cause a pressure build-up and induce microcracks. The hydrogen concentration that pre-charged under current density of 1 A/m^2 should be not high enough to form an sufficient internal pressure that can promote the initiation of microcracks, and no microcracks are observed on the fractured surfaces, as shown in Figures 8.9-8.11. As the hydrogen charging current density is increased from 1 to 10 A/m^2, the pressure build-up takes effect due to the high amount of pre-charged diffusible hydrogen in the steels, leading to microcracks occur, as observed in the 0.1W and 0.5W specimens (Figure 8.13a through d). Therefore, there is a critical hydrogen concentration that may cause the occurrence of microcracks. For the hydrogen-charged 0W and 1W specimens however, no microcracks are observed under the current density of 10 A/m^2. In the 0W specimen, the size of precipitates is coarse, and the interlamellar spacing of pearlite is large (see Section 3.5); the hydrogen trapping capability is expected to be low relative to the tungsten-added specimens. It is assumed that the accumulated hydrogen at the interfaces of precipitate/matrix and cementite/matrix is not high enough due to the coarse precipitated particles, large interlamellar spacing of pearlite, and as well as few dislocations which can act as hydrogen traps and transport during the notch tensile test. Therefore, microcracks are not initiated in the 0W specimen. A full cleavage fracture should have occurred before enough internal pressure is formed to cause microcracks due to the poor plasticity of the 1W specimen since a certain time is required for hydrogen accumulation at the interfaces of precipitate/matrix and cementite/matrix.

8.11. Mechanism of Hydrogen-Induced Embrittlement

Based on the aforementioned analysis and discussion, it can be concluded that microstructural features including dislocations, grain boundaries, precipitates and interlamellar spacing of pearlite may act as reversible and/or irreversible hydrogen traps in the steel to reduce the mobility of hydrogen towards regions of stress concentration ahead of the crack tip and thereby affect the susceptibility of HE and the fracture modes. When the hydrogen concentration in the traps increases above the critical value for crack initiation, the traps will act as crack initiation sites. Fine precipitates and low interlamellar spacing of pearlite can have positive and negative effects on the HE susceptibility of steels. In the one hand, fine precipitates and low interlamellar spacing of pearlite are more effective for hydrogen trapping, and can reduce the HE susceptibility [16]. In the other hand however, hydrogen trapping at fine precipitates and cementite may weaken the precipitate/matrix, cementite/matrix and Fe-Fe atomic bonding forces. When tungsten is added in the microalloyed steels, the size of precipitates is refined, and the interlamellar spacing is reduced; at the same time the dislocation density is increased. With an increase of dislocation density after tungsten addition, the amount of reversible hydrogen trapped at dislocations increases. The rate and amount of hydrogen accumulated at precipitate/matrix and cementite/matrix interfaces become high due to the losing of massive reversible hydrogen from the large amount of dislocations when they meet precipitates and cementite during notch tensile test. As a result, the hydrogen-induced embrittlement will be caused due to the aforementioned decohesion and internal pressure effects. The higher the tungsten content, the finer the precipitates, the lower the interlamellar spacing, and the higher the dislocation density are obtained, and correspondingly the more susceptible of HE is expected. The mechanism of hydrogen-induced embrittlement during notch tensile test is schematically illustrated in Figure 8.14.

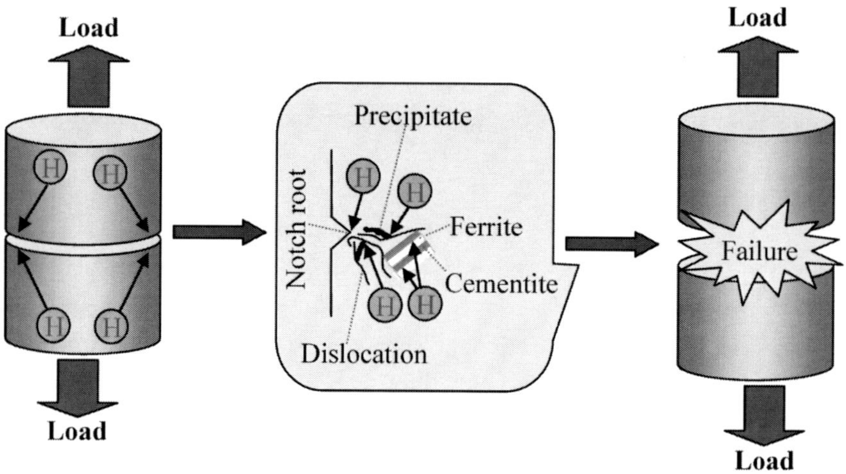

Figure 8.14. Schematic illustration of hydrogen-induced embrittlement during notch tensile test.

8.12. Summary

In this chapter, basic knowledge with respect to HE including categories of HE occurred in steels, factors affecting HE of steels and prevention strategies of HE of steels is introduced. Some existing investigations on the HE of microalloyed steels are briefly reviewed. The susceptibility of HE of microalloyed steels with different tungsten additions is assessed by SSRTs on circumferentially notched cylindrical specimens after cathodic hydrogen charging in a solution of "3 wt.% NaCl + 0.3 wt.% NH_4SCN" under the current densities of 1 to 20 A/m^2 for 48 h at room temperature. The average hydrogen concentration in a hydrogen-charged specimen is evaluated by means of TDS analysis. The hydrogen desorption behaviour, hydrogen desorption activation energy, notch tensile behaviour and fracture characterisation of these steels are systematically analysed. The mechanism of hydrogen-induced embrittlement is discussed.

The TDS analysis shows that an additional hydrogen desorption peak appears at the temperature of about 360 K for the 0W steel after hydrogen charging. The hydrogen desorption peak decreases gradually with the increase of tungsten, and the peak shifts to a higher temperature side as tungsten is added.

Tungsten addition decreases the amount of diffusible hydrogen in microalloyed steels under the same hydrogen charging current density. It is more difficult for hydrogen to permeate in the tungsten-containing steels during hydrogen charging relative to the steel without tungsten addition. Tungsten addition induces increased activation energy of hydrogen desorption.

Before hydrogen charging, the 0W, 0.1W and 0.5W steels exhibit a ductile fracture mode in the vicinity of notch root, and the area of ductile region decreases with increasing the tungsten content. A full cleavage fracture is observed in the uncharged 1W steel. Hydrogen charging does not show noticeable effect on the fracture mode in the central zone.

Hydrogen charging benefits the ductile to quasi-cleavage fracture transition, and thus promotes the occurrence of embrittlement in microalloyed steels. High current density (10 A/m^2 in this study) leads to the occurrence of microcracks along the margin of the notch root of the 0.1W and 0.5W steels, while no microcracks are found in the 0W and 1W steels. The change of microstructure after tungsten addition affects the occurrence of microcracks during notch tensile tests.

Microstructural features including dislocations, grain boundaries, precipitates and interlameller spacing of pearlite affect the HE behaviour and fracture modes of microalloyed steels. The hydrogen-induced embrittlement in the microalloyed steels studied is thought to be caused by the combined effects of decohesion and internal pressure in the presence of hydrogen.

REFERENCES

[1] L. Garverick, *Corrosion in the Petrochemical Industry*, ASM International, 1994.

[2] I.J. Park, K.H. Jeong, J.G. Jung, C.S. Lee, Y.K. Lee, The mechanism of enhanced resistance to the hydrogen delayed fracture in Al-added Fe-18Mn-0.6C twinning-induced plasticity steels, *International Journal of Hydrogen Energy*, 37 (2012) 9925-9932.

[3] J. Sojka, V. Vodárek, I. Schindler, C. Ly, M. Jérôme, P. Váňová, N. Ruscassier, A. Wenglorzová, Effect of hydrogen on the properties and fracture characteristics of TRIP 800 steels, *Corrosion Science*, 53 (2011) 2575-2581.

[4] A.M Elhoud, N.C. Renton, W.F. Deans, Hydrogen embrittlement of super duplex stainless steel in acid solution, *International Journal of Hydrogen Energy*, 35 (2010) 6455-6464.

[5] Y.S. Chun, J.S. Kim, K.T. Park, Y.K. Lee, C.S. Lee, Role of ε martensite in tensile properties and hydrogen degradation of high-Mn steels, *Materials Science and Engineering A*, 533 (2012) 87-95.

[6] J.S. Kim, Y.H. Lee, D.L. Lee, K.T. Park, C.S. Lee, Microstructural influences on hydrogen delayed fracture of high strength steels, *Materials Science and Engineering A*, 505 (2009) 105-110.

[7] L. Marchetti, E. Herms, P. Laghoutaris, J. Chêne, Hydrogen embrittlement susceptibility of tempered 9%Cr-1% Mo steel, *International Journal of Hydrogen Energy*, 36 (2011) 15880-15887.

[8] L.W. Tsay, M.Y. Chi, Y.F. Wu, J.K. Wu, D.-Y. Lin, Hydrogen embrittlement susceptibility and permeability of two ultra-high strength steels, *Corrosion Science*, 48 (2006) 1926-1938.

[9] T. Michler, M.P. Balogh, Hydrogen environment embrittlement of an ODS RAF steel - Role of irreversible hydrogen trap sites, *International Journal of Hydrogen Energy*, 35 (2010) 9746-9754.

[10] J.L. Albarran, L. Martinez, H.F. Lopez, Effect of heat treatment on the stress corrosion resistance of a microalloyed pipeline steel, *Corrosion Science*, 41 (1999) 1037-1049.

[11] ASTM F519, Standard Test Method for Mechanical Hydrogen Embrittlement Evaluation of Plating/Coating Processes and Service Environments, *ASTM International*, 2013.

[12] S.M. Lee, J.Y. Lee, The effect of the interface character of TiC particles on hydrogen trapping in steel, *Acta Metallurgica*, 35 (1987) 2659-2700.

[13] J. Zhao, Z. Jiang, C.S. Lee, Effect of tungsten on the hydrogen embrittlement behaviour of microalloyed steels, *Corrosion Science*, 82 (2014) 380-391.

[14] K. Kiuchi, R.B. McLellan, The solubility and diffusivity of hydrogen in well-annealed and deformed iron, *Acta Metallurgica*, 31 (1983) 961-984.

[15] W.Y. Choo, J.Y. Lee, Thermal analysis of trapped hydrogen in pure iron, *Metallurgical Transactions A*, 13A (1982) 135-140.

[16] H.J. Kang, J.S. Yoo, J.T. Park, S.T. Ahn, N. Kang, K.M. Cho, Effect of nano-carbide formation on hydrogen-delayed fracture for quenching and tempering steels during high-frequency induction heat treatment, *Materials Science and Engineering A*, 543 (2012) 6-11.

[17] J.Y. Lee, J. L. Lee, A trapping theory of hydrogen in pure iron, *Philosophical Magazine A*, 56 (1987) 293-309.
[18] J.L. Lee, J.Y. Lee, Identification of defects generated during cathodic charging in pure iron by thermal analysis technique, *Metallurgical Transactions A*, 16A (1985) 468-471.
[19] I. Maroef, Hydrogen trapping in ferritic steel weld metal, *International Materials Reviews*, 47(4) (2002) 191-223.
[20] M. Wang, E. Akiyama, K. Tsuzaki, Effect of hydrogen on the fracture behavior of high strength steel during slow strain rate test, *Corrosion Science,* 49 (2007) 4081-4097.
[21] R.A. Oriani, J.P. Hirth, M. Smialowski, *Hydrogen Degradation of Ferrous Alloys*, Noyes Publications, UK, 1985.
[22] K.Y. Lee, J.Y. Lee, D.R. Kim, A study of hydrogen-trapping phenomena in AISI 5160 spring steel, *Materials Science and Engineering*, 67 (1984) 213-220.
[23] J.L. Lee, J.Y. Lee, Hydrogen trapping in AISI 4340 steel, *Metal Science*, 17 (1983) 426-432.
[24] J.L. Lee, J.Y. Lee, The interaction of hydrogen with the interface of AI2O3 particles in iron, *Metallurgical Transactions A*, 17A (1986) 2183-2186.
[25] F.G. Wei, K. Tsuzaki, Quantitative analysis on hydrogen trapping of TiC particles in steel, *Metallurgical and Materials Transactions A*, 37A (2006) 331-353.
[26] G.M. Pressouyre, I.M. Bernstein, A quantitative analysis of hydrogen trapping, *Metallurgical Transactions A*, 9A (1978) 1571-1580.
[27] T. Michler, J Naumann. Microstructural aspects upon hydrogen environment embrittlement of various bcc steels, *International Journal of Hydrogen Energy*, 35 (2010) 821-832.
[28] B. Beidokhti, A. Dolati, A.H. Koukabi, Effects of alloying elements and microstructure on the susceptibility of the welded HSLA steel to hydrogen-induced cracking and sulphide stress cracking, *Materials Science and Engineering A*, 507 (2009) 167-173.
[29] NACE Standard TM0284-96, *Evaluation of Pipeline and Pressure Vessel Steels for Resistance to Hydrogen-Induced Cracking*, NACE International, Houston, 1996.
[30] W.K Kim, S.U. Koh, B.Y. Yang, K.Y. Kim, Effect of environmental and metallurgical factors on hydrogen induced cracking of HSLA steels, *Corrosion Science*, 50 (2008) 3336-3342.

[31] NACE Standard TM0284-02, *Evaluation of Pipeline and Pressure Vessel Steels for Resistance to Hydrogen-Induced Cracking*, NACE International, Houston, 2002.

[32] S.J. Kim, H.G. Jung, K.Y. Kim, Effect of post-weld heat treatment on hydrogen-assisted cracking behaviour of high-strength process pipe steel in a sour environment, *Scripta Materialia*, 67 (2012) 895-898.

[33] NACE Standard TM0284-2003, *Evaluation of Pipeline and Pressure Vessel Steels for Resistance to Hydrogen-Induced Cracking*, NACE International, Houston, 2003.

[34] N.E. Nanninga, Y.S. Levy, E.S. Drexler, R.T. Condon, A.E. Stevenson, A.J. Slifka, Comparison of hydrogen embrittlement in three pipeline steels in high pressure gaseous hydrogen environments, *Corrosion Science*, 59 (2012) 1-9.

[35] T. Alp, F. I. Iskanderani, A.H. Zahed, Hydrogen effects in a dual-phase microalloy steel, *Journal of Materials Science*, 26 (1991) 5644-5654.

[36] M. Opiela, Hydrogen embrittlement of welded joints for the heat-treatable XABO 960 steel heavy plates, *Journal of Achievements in Materials and Manufacturing Engineering*, 38 (2010) 41-48.

[37] A.P. Moon, R. Balasubramaniam, B. Panda, Hydrogen embrittlement of microalloyed rail steels, *Materials Science and Engineering A*, 527 (2010) 3259-3263.

[38] K. Banerjee, U.K. Chatterjee, Hydrogen embrittlement of an HSLA 80 steel in sea water under cathodic charging conditions, *Materials Science and Technology*, 16 (2000) 517-523.

[39] L. Vehovar, Hydrogen embrittlement of microalloyed structural steels, *Materials and Corrosion*, 45 (1994) 349-354.

[40] K. Ravi, V. Ramaswamy, T.K.G. Namboodhiri, Effect of molybdenum on the resistance to H2S of high sulphur microalloyed steels, *Materials Science and Engineering A*, 169 (1993) 111-118.

[41] D. Hardie, Su'e Liu, The effect of stress concentration on hydrogen embrittlement of a low alloy steel, *Corrosion Science*, 38 (1996) 721-733.

[42] L.W. Tsay, W.C. Lee, W.C. Luu, J.K. Wu, Effect of hydrogen environment on the notched tensile properties of T-250 maraging steel annealed by laser treatment, *Corrosion Science*, 44 (2002) 1311-1327.

[43] E. Proverbio, P. Longo, Sub critical crack growth in hydrogen assisted cracking of cold drawn eutectoid steel, *Corrosion Science*, 49 (2007) 2421-2435.

[44] M. Wang, E. Akiyama, K. Tsuzaki, Effect of hydrogen and stress concentration on the notch tensile strength of AISI 4135 steel, *Materials Science and Engineering A*, 398 (2005) 37-46.

[45] S.H. Wang, W.C. Luu, K.F. Ho, J.K. Wu, Hydrogen permeation in a submerged arc weldment of TMCP steel, *Materials Chemistry and Physics*, 77 (2002) 447-454.

[46] W.C. Luu, J.K. Wu, The influence of microstructure on hydrogen transport in carbon steels, *Corrosion Science*, 38 (1996) 239-245.

[47] G.T. Park, S.U. Koh, H.G. Jung, K.Y. Kim, Effect of microstructure on the hydrogen trapping efficiency and hydrogen induced cracking of linepipe steel, *Corrosion Science*, 50 (2008) 1865-1871.

[48] H.E. Kissinger, Reaction kinetics in differential thermal analysis, *Analytical Chemistry*, 29 (1957) 1702-1706.

[49] F.G. Wei, T. Hara, K. Tsuzaki, Precise determination of the activation energy for desorption of hydrogen in two Ti-added steels by a single thermal-desorption spectrum, *Metallurgical and Materials Transactions B*, 35 (2004) 587-597.

[50] G. Krauss, *Steels: Processing, Structure, and Performance*, ASM International, Ohio, 2005.

[51] S. Taketomi, R. Matsumoto, N. Miyazaki, Atomistic study of the effect of hydrogen on dislocation emission from a mode II crack tip in alpha iron, *International Journal of Mechanical Science*, 52 (2010) 334-338.

[52] P.J. Ferreira, I.M. Robertson, H.K. Birnbaum, Hydrogen effects on the interaction between dislocations, *Acta Materialia*, 46 (1998) 1749-1757.

[53] H. Lü, M. Li, T. Zhang, W. Chu, Hydrogen-enhanced dislocation emission, motion and nucleation of hydrogen-induced cracking for steel, *Science in China (Series E)*, 40 (1997) 530-538.

[54] J. Zhao, T. Lee, J.H. Lee, Z. Jiang, C.S. Lee, Effects of tungsten addition on the microstructure and mechanical properties of microalloyed forging steels, *Metallurgical and Materials Transactions A*, 44 (2013) 3511-3523.

[55] E. Ramírez, J.G. González-Rodriguez, A. Torres-Islas, S. Serna, B. Campillo, G. Dominguez-Patiño, J.A. Juárez-Islas, Effect of microstructure on the sulphide stress cracking susceptibility of a high strength pipeline steel, *Corrosion Science*, 50 (2008) 3534-3541.

[56] M.C. Zhao, M. Liu, A. Atrens, Y.Y. Shan, K. Yang, Effect of applied stress and microstructure on sulfide stress cracking resistance of pipeline steels subject to hydrogen sulfide, *Materials Science and Engineering A*, 478 (2008) 43-47.

[57] T. Michler, M.P. Balogh, Hydrogen environment embrittlement of an ODS RAF steel - Role of irreversible hydrogen trap sites, *International Journal of Hydrogen Energy*, 35 (2010) 9746-9754.
[58] R.A. Oriani, Hydrogen-the versatile embrittler, *Corrosion*, 43 (1987) 390-397.
[59] A. Torres-Islas, S. Serna, B. Campillo, J. Colin, A. Molina, Hydrogen embrittlement behavior on microalloyed pipeline steel in NS-4 solution, *International Journal of Electrochemical Science*, 8 (2013) 7608-7624.
[60] L.W. Tsay, H.L. Lu, C. Chen, The effect of grain size and aging on hydrogen embrittlement of a maraging steel, *Corrosion Science*, 50 (2009) 2506-2511.
[61] M. Imade, S. Fukuyama, L. Zhang, M. Wen, K. Yokogawa, Hydrogen environment embrittlement of SCM440 steel in high pressure hydrogen at room temperature, *Journal of The Japan Institute of Metals*, 69 (2005) 190-193.
[62] H.K. Birnbaum, P. Sofronis, Hydrogen-enhanced localized plasticity - a mechanism for hydrogen-related fracture, *Materials Science and Engineering A*, 176 (1994) 191-202.
[63] M. Nagumo, Function of hydrogen in embrittlement of high-strength steels, *ISIJ International*, 41 (2001) 590-598.
[64] J.P. Hirth, Effects of hydrogen on the properties of iron and steel, *Metallurgical Transactions A*, 11A (1980) 861-890.
[65] D. Pérez Escobar, T. Depover, E. Wallaert, L. Duprez, M. Verhaege, K. Verbeken, Thermal desorption spectroscopy study of the interaction between hydrogen and different microstructural constituents in lab cast Fe-C alloys, *Corrosion Science*, 65 (2012) 199-208.
[66] Y. Qi, H. Luo, S. Zheng, C. Chen, D. Wang, Effect of immersion time on the hydrogen content and tensile properties of A350LF2 steel exposed to hydrogen sulphide environments, *Corrosion Science*, 69 (2013) 164-174.
[67] R. Wang, Effects of hydrogen on the fracture toughness of a X70 pipeline steel, *Corrosion Science*, 51 (2009) 2803-2810.
[68] N.R. Raykar, R.K. Singh Raman, S.K. Maiti, L. Choudhary, Investigation of hydrogen assisted cracking of a high strength steel using circumferentially notched tensile test, *Materials Science and Engineering A*, 547 (2012) 86-92.

Chapter 9

CONCLUSION

As an alloying element with growing scientific and industrial interests, tungsten has been putting great attention due to its positive roles in microalloyed steels. This book provides up-to-date, informative, readable and self-contained information for the metallurgists, engineers, materials scientists, academics and higher degree research students who make, use, study and design with steels and tungsten throughout the world by introducing the latest research findings and achievements of tungsten alloying in microalloyed steels, and considering the metallurgical basis for steel research.

This book has systematically described the functions of tungsten alloying in microalloyed steels. Many aspects that need to be assessed for microalloyed steels when they are used in practice have been addressed. These aspects include the background knowledge of tungsten and microalloyed steels, and the effects of tungsten on the phase transformation, precipitation, microstructural evolution, mechanical properties, optimal heat treatment conditions, corrosion resistance and hydrogen embrittlement. In each chapter, relevant basic principles and concepts are firstly introduced to help readers understand well the contents being presented, and the main findings of the roles of tungsten alloying in microalloyed steels are summarised at the end of each chapter.

Tungsten alloying in microalloyed steels is a rear and fruitful research field, and only limited published work is available with related to this topic. A lot of work is still needed to do in future to explore the best use of tungsten as an alloying element in microalloyed steels with the purposes of obtaining the optimal tungsten content that results in the best overall performance as well as the optimal processing conditions for producing defect-free products. In this book, we describe only the functions of tungsten alloying in hot forged microalloyed steels. The following work is still valuable to be done in the future research:

(1) Roles of tungsten alloying in steelmaking and rolling of microalloyed steels.
(2) Determination of optimal tungsten contents in different grades of microalloyed steels.
(3) Design of optimal processes for steelmaking, forging and rolling to produce defect-free microalloyed steels with different tungsten additions.
(4) Effects of tungsten addition on the weldability of microalloyed steels.
(5) Formability of the tungsten-containing microalloyed steels.
(6) Practical applications of tungsten-containing microalloyed steels.

We hope this book will be a turning-point for the development of new microalloyed steels with improved quality and properties to make them applicable to various application environments.

INDEX

A

B

C

I

J

K

L

M

N

O

P

Q

R

S

U

V

W

X

Y

Z

α

ε